Sustainable Materials and Technology

Sustainable Materials and Technology (SMT) book series publishes research monographs (both edited and authored volumes) showcasing the latest developments in the field and comprehensively covering topics such as:

- Recycling of waste into useful material and their energy applications
- Catalytic action of Nano oxides for efficient carbon reforming process
- Sustainable technologies for plastic transformation
- Bifunctional nanoparticles for sustainable water splitting applications
- Sustainable dying and printing
- New materials from waste
- Sustainable Manure Management and Technology: Potentials, Uses and limitations
- Sustainable Mechanical Engineering Approach
- Sustainable biochemistry for the improvement of health
- Sustainable development of Mechanical recycling of automotive components
- Sustainable-waste recycling and conversion in useful materials for different applications
- Sustainable development of inexpensive Nano-photo catalysts
- Sustainable development of recycling of discarded lithium ion batteries
- Modern sustainable cement and concrete
- Sustainable adsorbent for hazardous removal
- Sustainable superior electromagnetic shielding materials
- Excellent sustainable nanostructured materials for energy storage device
- Sustainable development of heavy metal detoxification from water
- Carbon dioxide utilization for sustainable energy
- Sustainable development in green syntheses of materials
- Environment friendly and sustainable cloth for garments application
- Sustainable design and application of eco-materials
- Nanoparticles for sustainable environment applications
- Sustainable remediation of industrial contaminated water towards potential industrial applications
- Biomaterials for sustainable bioremediations

Anish Khan · Abdullah Asiri · Showkat Bhawani
Editors

Waste to Biofuel Technology

Future Energy

Springer

Editors
Anish Khan
Center of Excellence for Advanced Materials Research
King Abdulaziz University
Jeddah, Saudi Arabia

Abdullah Asiri
Chemistry Department, Faculty of Science
King Abdulaziz University
Jeddah, Saudi Arabia

Showkat Bhawani
Faculty of Resource Science and Technology
Universiti Malaysia Sarawak
Kota Samarahan, Malaysia

ISSN 2731-0426 ISSN 2731-0434 (electronic)
Sustainable Materials and Technology
ISBN 978-981-97-4560-9 ISBN 978-981-97-4561-6 (eBook)
https://doi.org/10.1007/978-981-97-4561-6

This Springer imprint is published by the registered company Springer Nature Singapore Pte Ltd.
The registered company address is: 152 Beach Road, #21-01/04 Gateway East, Singapore 189721, Singapore

Contents

Valorization of Palm Biomass Wastes for Biodiesel Production

João H. C. Wancura, Maicon S. N. dos Santos, Carolina E. D. Oro, J. Vladimir de Oliveira, and Marcus V. Tres

Abstract The intensive and exacerbated use of fossil fuels has alarmed the global industrial system and has led to the adoption of sustainable alternatives aimed at minimizing the serious environmental impacts caused by the exploitation of these resources. One of the primordial scientific topics in recent years is the investigation for potential safe alternative fuels as a path to substitute the current energy exploitation. Appropriately, the use of plant biomass as potential agents in the synthesis of biofuels has been a highly viable and sustainable alternative, mainly from an environmental and economic outlook. Therefore, the palm oil and palm oil-based industries have been identified as prominent sources of bioenergy through the conversion of palm biomass into products with high added value. Consequently, the diversity of palm biomasses has indicated encouraging potentiality for application in a large-scale biofuel production chain, such as mesocarp fibers, trunk, frond, and kernel feedstocks. Accordingly, this study described the main application of palm biomass for biodiesel production. Contextually, the chapter presented valuable information on the main technological strategies applied for biodiesel preparation, such as pyrolysis, transesterification, and enzymatic and supercritical techniques. Additionally, the study comprehended important natural sources widely explored as potential products for biodiesel synthesis, highlighting the current scientific research on oil

J. H. C. Wancura
Laboratory of Biomass and Biofuels (L2B), Federal University of Santa Maria (UFSM), Santa Maria, RS, Brazil

M. S. N. dos Santos · M. V. Tres (✉)
Laboratory of Agroindustrial Processes Engineering (LAPE), Federal University of Santa Maria (UFSM), Cachoeira Do Sul, RS, Brazil
e-mail: marcus.tres@ufsm.br

C. E. D. Oro
Department of Food Engineering, Regional Integrated University of Upper Uruguai and Missions (URI), Erechim, RS, Brazil

J. V. de Oliveira
Department of Chemical and Food Engineering, Federal University of Santa Catarina (UFSC), Florianópolis, SC, Brazil
e-mail: jose.vladimir@ufsc.br

A. Khan et al. (eds.), *Waste to Biofuel Technology*, Sustainable Materials and Technology, https://doi.org/10.1007/978-981-97-4561-6_1

palm biomass. Moreover, the main characteristics of these biomasses as an encouraging feedstock option for biodiesel preparation were provided. Finally, conclusions and relevant future perspectives on the advancement of the oil palm feedstocks for biodiesel preparation and production were determined.

Keywords Bio-oil · Bioenergy · Antioxidants · Biorefinery approach · Oil palm biomasses · Plant wastes · Potential feedstocks · Sustainable agriculture · Transesterification · Vegetable add-value by-products

1 Introduction

Biodiesel (fatty acid methyl esters—FAME) is a renewable, non-toxic, and biodegradable biofuel usually obtained through a reaction involving oleaginous sources and short-chain alcohol with (or not) a catalyst [1]. This biofuel is considered an environmentally friendly energy source since it can offer a reduction in the amounts of direct and indirect greenhouse gases (like particulate matter, sulfur, hydrocarbonss and mainly carbon dioxide, and monoxide), protecting the environment [2]. Biodiesel also can be employed in diesel engines with minimal or no mechanical modification, being typically mixed with petrodiesel in volumetric proportions stipulated in government policies according to the country.

The main route to biodiesel production on industrial demands is through a homogeneous alkaline transesterification reaction, where at 64 °C, with 0.5 wt% of sodium methoxide, oil to methanol molar ratio of 1.0:6.0 and 1.0 h reaction, it is possible to achieve biodiesel yields above 97% [3, 4]. Within this perspective, diverse technological methods, catalyzed by different chemical compounds (heterogeneous alkaline and,homogeneous and heterogeneous acid) or enzymatic have been considered the target of several studies published recently for biofuel preparation [5, 6].

Distinct oils (or lipids) from either plant or animal can become a raw material for biodiesel synthesis. Typically, potential feedstocks used for biodiesel preparation are soybean, rapeseed, and palm oil, depending on the region where the determined oleaginous source has a predominance of cultivation [7]. These feedstocks can be categorized as the first, second, third, and fourth generation. First-generation raw materials are principally extracted from edible crops or biomass (palm, soybean oil, rapeseed, and sunflower, for example). Second-generation raw materials for biodiesel production originated from non-edible sources, such as WCO, grease, animal fat, and non-edible plant oils. Third-generation feedstocks originated from algal biomass. Feedstock destined for biodiesel of fourth-generation can be synthesized through one or a junction of three resources: specific photosynthetic microorganisms; a combination of photovoltaics production and microbial fuel; and (or) production of the artificial cell [7, 8].

Among the different biodiesels, those synthetized by palm residues have demonstrated themselves as the most dominant biofuel by virtue of the largest impact worldwide (35%), lowest cost (660 USD·ton^{-1}), higher oil content (5,000 kg oil·ha^{-1}), and yield of production (4.2 Mton·ha^{-1}) since it needs less sunlight to yield one unit of oil per hectare [9]. Moreover, the cultivation of palm trees can give fruit from fourth-year forward and can be reaped until 50 years [10]. The worldwide market of palm oil is developing gradually and it is projected to achieve a value market of US$ 92.8 billion in 2021 [9]. According to Ong et al. [11], in 2050, it is projected that the demand for palm oil will reach a massive amount of 240 Mton.

The application of edible oils, such as palm oil, for biodiesel synthesis, could result in growing food prices considering the increased demand for fuels and food products [12]. In terms of production costs, it is important to cite that the employment of edible oils as suitable raw material can represent 70 to 80% of the total, influencing the feasibility of determining projects [13]. Moreover, although non-edible oils such as WCO can directly be converted to biodiesel, these raw materials are not sufficiently available for large production and also can have their performance affected in cold weather [14]. In this scenario, the operation with alternative raw materials for biodiesel preparation is an interesting way to circumvent this type of situation.

Among the numerous unusual raw materials investigated for biodiesel preparation, palm wastes biomass has been the focus of different research recently [15–17). This situation may be associated with the current evolution of the palm oil industry and an expectation to result in a substantial increase in the quantities of solid biomass wastes [12]. Generally, one hundred tons of fresh palm bunches produce fifty tons of biowastes, particularly empty fruit bunches [18]. Nevertheless, is important to highlight that oil palm tree biomass comprises not only empty fruit bunch but also trunks, oil palm fronds, palm shell kernel, and mesocarp fiber, all with the potential for oil extraction (see Fig. 1). Therefore, with a promising demand for palm oil, it is projected that a generation of oil palm by-products will grow with the increase of this demand. These factors represent an excellent opportunity for eco-friendly biodiesel production employing oil palm biowastes as added-value raw material.

Furthermore, it is appropriate to establish a quantitative aspect of scientific studies on the applications of palm oil and palm raw materials as valuable resources for biodiesel production, particularly in the industrial context. From an academic/scientific background, the application of feedstocks for biofuel production has been largely identified as an advanced and highly sustainable alternative. Accordingly, the growth of scientific publications recently reported on the application of oil palm feedstocks is interesting and a pertinent factor to be discussed in this chapter. In this context, advanced research was performed on the scientific-based Web of Science® platform. To perform a particular investigation of the scientific articles published in the exploration of the topic of this study, the following keywords were assumed: *Oil palm biodiesel biomasses*; *oil palm kernel/oil palm mesocarp/oil palm trunk/oil palm frond* (Fig. 2).

The scientific investigation on the employment of oil palm feedstocks as potential sources for biodiesel synthesis indicated an expressive growth in the productive scenario in the last 10 years. Data acquired by the Web of Science® platform showed a large increase in addressing kernel, mesocarp, trunk, and frond biomass since 2011. 140 manuscripts were published in 2011 combining the different oil palm feedstocks. In 2020, 311 studies were counted. This growth was mainly driven

Fig. 1 Wastes generated from a palm trees

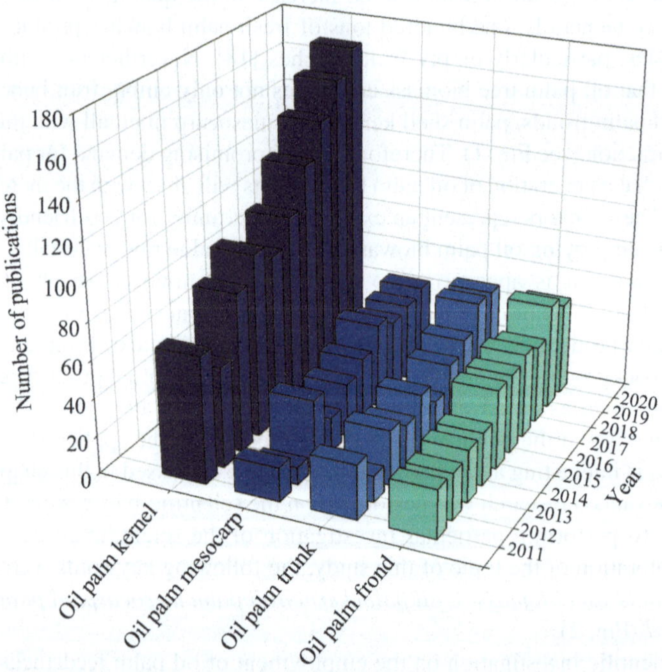

Fig. 2 Research on the dimension and frequency of publication of scientific articles on the application of oil palm feedstocks for biofuel production arranged by year from 2011 to 2020 according to the Web of Science® scientific-based platform with the following keywords: *Oil palm biodiesel biomasses*; *oil palm kernel/oil palm mesocarp/oil palm trunk/oil palm frond*

by the exploitation of the oil palm kernel resource. A range of technological strategies and the potentialization of the activity of this material in the biodiesel synthesis stages have been adopted recently. This panorama is strongly demonstrated by the resistant advance of the number of studies that investigate this biomass in biodiesel production processes. In 2011, 67 studies were published in scientific journals. In 2020, there were 169 manuscripts. Appropriately, the verified scenario expresses the importance of the magnitude of the pillars of economic, social, and environmental sustainability in the biodiesel production chain and reports the utilization of oil palm feedstocks as promising and competitive resources that have been largely potentialized for application in a global perspective.

Appropriately, the motivation of this chapter was to perform a comprehensive approach to value the palm oil biomass as feedstock to be applied to biodiesel production. For this proposal, the main technologies and wastes applied to biodiesel preparation, a discussion of the main studies published in recent years where palm oil biomass wastes to synthesize biodiesel was employed, as well as future perspectives related to the topic are discussed.

2 Technologies Applied to Biodiesel Preparation

Biodiesel can be obtained from vegetable oils or animal fat through the process of transesterification, thermal cracking (pyrolysis), or supercritical technique with or without the use of different catalysts, including homogeneous or heterogeneous chemical catalysts and enzymes [19]. Some details of the main technologies applied for the production of biodiesel are presented below.

2.1 Pyrolysis

Pyrolysis or thermal cracking is a process that causes the breakdown of molecules by heating at high temperatures in the absence of air or oxygen at temperatures above 450 °C. The final product consists of a mixture of chemical compounds with properties very similar to those of petroleum diesel. Pyrolysis processes employ esters of an alkyl chain attached to the ester group. In these processes, the main changes in the reaction are the number of double bonds and carbon atoms. This process can employ catalysts to break chemical bonds and generate smaller molecules [20–22]. The fuel obtained by cracking oils and fats is not considered biodiesel under international nomenclature. For this reason, it is called by the terms ecodiesel, cracked biodiesel, and bio-oil [23, 24].

Two distinct and successive steps occur in the thermal cracking process. First, the formation of carboxylic acids occurs by breaking carbon and oxygen bonds between the glyceride part and the rest of the triglyceride carbon chain. Then, the carboxylic acids formed in the initial step are deoxygenated through the routes of decarboxylation and decarbonylation [25].

In order to eliminate oxygenated products, some catalysts can be used to change the selectivity of pyrolysis products. The most used catalysts are aluminum oxide, silicon oxides, molybdenum oxides, nickel supported on alumina, acidic zeolites, phosphoric acid supported on silica, alumina doped with metallic oxides (tin and zinc), and several others based on silica [19, 23, 26]. Alumina, on the other hand, as it is not very selective, reduces the final acidity of the products and should be used in secondary cracking. Furthermore, pyrolysis is widely used in biomass to produce a combination of gases, biochar, and bio-oil [27, 28].

2.2 *Transesterification*

2.2.1 Alkaline Route

The most used technological route for the production of biodiesel in the world is the technological route of methyl transesterification, in which vegetable oils and/or animal fat are mixed with methanol and associated with a basic catalyst, biodiesel is produced. The alkaline route generates large amounts of glycerin, which must be separated from the final product. However, the purification steps of the final products (biodiesel and glycerin) in this process require the recovery or neutralization of the catalyst and the removal of the saponification products that are generated in the reaction. Another drawback of base-catalyzed transesterification is water intolerance. The negative effect occurs because water consumes the catalyst and reduces its efficiency in the process [29].

The use of catalysts in the alkaline route, such as sodium hydroxide (NaOH) or potassium hydroxide (KOH), presents a fast reaction rate reaction. Biodiesel obtained by the alkaline route can be generated by homogeneous and heterogeneous processes. The transesterification of vegetable oils and/or animal fats in a homogeneous alkaline medium is still the most used method for the production of biodiesel [29]. However, some reaction inconveniences should be highlighted in the homogeneous reaction, such as the difficulty in separating the final product, large volumes of effluents, and the impossibility of reusing catalysts [19]. As for the process with the application of enzymatic catalysts, the use of heterogeneous catalysts overcomes these inconveniences. The interest is in obtaining a stable catalyst, with ease of separation from the reaction medium and that presents the possibility of being used in different operational recycles [30, 31], and the so-called heterogeneous reaction is performed. However, the heterogeneous process has a slow conversion rate compared to the homogeneous catalyst [12].

In this reaction route, special attention to the raw material and the catalyst must be given. The use of used vegetable oils, as in the case of household or restaurant frying oil residue, can form soap when reacting with the catalyst. Furthermore, the yield of biodiesel is decreased due to the presence of contaminants such as free fatty acids and water in this used oil [19]. In this case, the use of other catalysts is recommended, such as CaO [19].

2.2.2 Acid Technique

Transesterification can also be carried out in an acid medium. One of the advantages over the use of alkaline catalysts is that the amount of free fatty acids in the medium is not relevant, as they can be esterified in the same reaction vessel. Therefore, it is possible to use raw materials with lower added value in the process. However, acidic transesterification requires more energetic conditions compared to the alkaline process. In acidic esterification, usually sulfuric and sulfonic acids are used [32, 33].

2.3 *Enzymatic Technique*

The enzymatic synthesis of biodiesel presents some advantageous characteristics in relation to other production processes. Enzymatic synthesis employs mild temperature conditions, there is no generation of effluents, and there is no saponification due to the competence of enzymes in converting reagents into products. For this type of enzymatic reaction, lipases are used in their free or immobilized form [6, 34]. Lipases (EC 3.1.1.3) are classified as hydrolases and act on ester bonds present in acylglycerols, releasing fatty acids and glycerol, being a special class of esterases [35].

Lipases can be used in the reaction in their free or immobilized form. The immobilized form allows the separation and reuse of the immobilized in successive cycles. Therefore, it is necessary to choose a support that presents low cost and quick synthesis for the process to be viable [36, 37]. Wancura et al. [38] used the commercial Callera™ Trans L free lipase in the hydrolysis-esterification reaction to produce biodiesel with soybean oil and methanol and achieved a yield of 96.9% in 24 h. Furthermore, Wancura et al. [34] showed that it is possible to achieve a yield of 97.1% in 8 h through a two-step hydroesterification reaction system using NS40116 lipase. The result presented by the authors shows that the use of lipase in its liquid form presents a high biodiesel conversion rate.

The enzymatic process is considered a "green" process and has many advantages; however, it has lower reaction rates compared to the alkaline process. Another point worth mentioning is the high cost of enzymes, which can affect the price of the process [38].

2.4 *Supercritical Technique*

Biodiesel production by supercritical route has been studied in recent years. The supercritical fluid condition is reached when the temperature and pressure of the reaction system are above the critical point of the components. The supercritical fluid can actively participate in the reaction or promote the solubilization of system components (homogeneous phase condition). The use of a supercritical fluid as the reaction medium increases the selectivity of the reaction, provides high yields, and improves the separation process of the system components [39, 40]. The raw materials commonly used for the supercritical process are methanol or ethanol and triglycerides [5].

In this process, the reaction yield and speed are high and the product purification steps are simplified compared to other processes since it does not use homogeneous catalysts and is not affected by the presence of water and free fatty acids. These steps would basically be the separation of the biodiesel phase from the glycerin phase through decantation [5, 41, 42].

However, some aspects must be considered for the production of biodiesel under supercritical conditions on a commercial scale. Due to the high temperatures and pressures required, the equipment needed for the production of biodiesel on a commercial scale would be expensive. It is necessary to carry out the complete energy balance of this new biodiesel production route, in order to establish in which situations, it could be more advantageous than the conventional route [39]. Based on the assumption, Table 1 presents some results from the literature on obtaining biodiesel from palm oil using the different techniques described.

3 Residues as Feedstock for Biodiesel Production

The use of alternative feedstocks for the production of biodiesel is an attractive and constantly expanding area among researchers. The direct application of feedstocks in the bioenergy industry emerges as a highly sustainable and economically viable strategy. Converting a range of by-products into value-added products generates many precursors for wide use and, at any level, the role of these biomaterials plays a fundamental role in enhancing feedstocks as competitive products and reversing the pollution caused by the inadequate disposal of these elements in the environment. Appropriately, this section reports the use of different by-products as valuable components in the biodiesel production chain.

Table 1 Methods to formulating bioactive compounds for phytopathogenic control in agriculture

Method applied	Feedstock	Process conditions	Biodiesel Yield (%)	Reference
Pyrolysis	Palm kernel shell	500–504.3 °C, 30 mL min^{-1} flow rate, 18.9–20 °C min^{-1} heating rate, and 30 min	36.9 (PKS-biochar)	[28]
Alkaline Route	Palm oil	80 °C, ethanol to oil molar ratio of 12:1, 5 wt% of SrO (total basic site of 3,725 $\mu mol \cdot g^{-1}$), 3 h, and 500 rpm	98.2	[43]
	Palm oil	80 °C, methanol to oil molar ratio of 15:1, 3 wt% of K_2O-CaO derived from chicken eggshell, and 2 h	87.9	[44]
	Palm oil	65 °C, methanol to oil molar of 12:1, 2 wt% of zeolite derived catalyst (Li_3AlSiO_5), and 46 min	99.7	[45]
	Palm oil	66.1 °C, methanol to oil molar ratio of 18.5:1, catalyst (Ca-Ce-Zn-Al_2O_3) load of 8.2 wt%, and 3 h	99.4	[46]
	Palm oil mill effluent	60 °C, methanol to oil molar ratio of 1:10, 1.5 wt% of KOH, and 2.5 h	93.0	[47]
Acid Route	Waste palm oil	130 °C, methanol to oil molar ratio of 12:1, 10 wt% of heterogeneous acid catalyst (derived from coconut coir husk), 3 h, and 500 rpm	89.8	[48]
	Palm oil	100 °C, methanol to oil molar ratio of 20:1, 10 wt% of sulfonated graphene (GR-SO_3H), and 14 h	98.0	[49]
	Palm oil	70 °C, methanol to oil molar ratio of 20:1, catalyst (MgO/CP) load of 4 wt%, 2.5 h, and 1250 rpm	97.6	[50]
Enzymatic Catalysis	Sludge palm oil	45 °C, methanol to oil molar ratio of 5:1, 0.2 wt% of lipase in liquid formulation (enzymatic activity of 9,600 $IU \cdot mL^{-1}$), 2.0 wt% of water, 24 h, and 750 rpm	94.0	[51]

(continued)

Table 1 (continued)

Method applied	Feedstock	Process conditions	Biodiesel Yield (%)	Reference
	Palm oil mill effluent	40 °C, methanol to oil molar ratio of 6:1, 2 g of immobilized *C. rugosa* lipase in PVA-alginate-sulfate beads, 5 h, and 300 rpm	65.0	[52]
	Palm oil	47 °C, methanol to oil molar ratio of 2.3:1, 15 wt% of combined enzyme load (between immobilized lipases *C. antarctica* B and *T. lanuginosa* at 2.1:1), 45 wt% of *t*-butanol, and 24 h	94.0	[53]
	Waste frying palm oil	30 °C, methanol to oil molar ratio of 7:1, 5.5 wt% of free lipase *C. antarctica* A, 16.6 wt% of water, 22 h, and 400 rpm	94.6	[54]
Supercritical Technique	Refined palm oil	360 °C, 8 MPa, *iso*-propanol to oil weight ratio of 0.2:1 (co-reactant), ethanol to oil molar ratio of 25:1, and 4.6 min	97.0	[55]

3.1 Animal Fat Waste

Animal by-products have been considered interesting resources for biodiesel production, particularly in bioenergy industries. Animal fats are mostly composed of concentrations of triglycerides, proteins, water, and a range of minerals [56]. The significant amounts of slaughtered animals over the years indicated the necessity to direct waste to operations that add value to these by-products in a large-scale perspective. The animal fat waste comprises tallows from the adipose tissue of cattle, sheep tallow, pork lard, chicken fat, and animal blood, skin, viscera, and trimmings [57]. Moreover, contemporary studies have detailed the application of animal fat waste and its by-products as valuable options, such as chicken feather meal [58], chicken skin, fat, and complex tissues [59], beef tallow [60], sheep fat [61], etc.

Depending on the occurrence of fatty acid contents in its composition, pretreatments are necessary in order to eliminate these components and optimize the obtainment of pure biofuel. The fatty acids in greatest quantity are saturated fatty acids (palmitic and stearic acids), monounsaturated fatty acids (oleic acid), and polyunsaturated fatty acids (linoleic and arachidonic acids) [57]. The characterization and properties of fatty acids associated with triacylglycerols is a fundamental factor in determining essential parameters, such as viscosity, iodine number, and oxidative stability [62]. The determination of this is decisive for the quality of the desired product, as well as being a limiting factor to the industrial process viability. Therefore, it is essential that operating and animal feedstock costs are considerably lower than the prices of competitive feedstocks such as plant-based oils [63].

Furthermore, when compared to plant-based biodiesel, the animal fat-based product has a lower content of unsaturated fatty acids, a lower proportion of cetane, a higher calorific value, and a low oxidative stability [64]. Moreover, it has high viscosity (up to 40 mm^2/s) and a lower flash point [62]. Nevertheless, the conversion of animal fat into value-added products drastically minimizes biodiesel costs and reports this alternative as an interesting option when compared to the fossil fuel market [65].

3.2 Waste Cooking Oil (WCO)

The WCO produced after the frying procedure underwent polymerization, oxidation, and hydrolysis reactions [66]. This operation results in the loss of the nutritional quality of the product, additionally to the generation of degradative compounds. Approximately 15 million tons of WCO are produced annually worldwide [67]. A massive fraction is directed to the environment, drastically affecting the sources of natural resources, stability of human and animal health, and preventing the possibility of adding value to the generated by-product. This scenario makes this biomass an attractive by-product in the bioenergy context.

The use of WCO leads to the minimization of waste generation and costs, acting as an extremely attractive action under the economic, environmental, and waste management approaches [68]. Furthermore, WCO can be reused in the development of sustainable commodities, reducing the environmental pressure brought about by targeting the environment and returning as a competitive product in the commercial routine [67]. A physicochemical comparison with different sources of biodiesel and diesel reported that the WCO had higher kinematic viscosity, flash point, and density [66]. The current application of this feedstock has resulted from the conversion of WCO fatty acids in methyl esters, the key components of biodiesel [69].

The transesterification of triglycerides, largely present in WCO residues for the synthesis of methyl/ethyl esters, is the conventional method of converting biomass to biodiesel [70]. However, the transesterification process results in some problems, since the generated by-products require supplementary processes to become useful materials, making the economic and sustainable footprint of the biodiesel synthesis chain unfeasible [70]. This scenario resulted in the adoption of strategies that optimize the process and facilitate the achievement of the final product. Recently, the exploration of catalysts has been encouraged, mainly due to the physicochemical characterization of these materials, additionally to their easy separation and reuse [71]. A range of works aimed at the application of this alternative in WCO residues has been noticed [72–74].

4 Palm Biomass Wastes as Potential Raw Material for Biodiesel Synthesis

Oil palm biomasses are successful bioenergy sources and the abundance of these materials has been reported as a potential scenario for processing industries. Oil palm kernel shells, mesocarp fiber, trunk, and frond wastes have been positively explored by a range of industries and scientific studies [75]. With different concentrations of compounds, palm wastes are essentially lignocellulosic materials, consisting of significant concentrations of cellulose, hemicellulose, and lignin. Accordingly, Table 2 reports the lignocellulosic contents of the different oil palm biomass. This panorama confers high resilience to the fibrous structures of palm oil since the resistant structure of lignin involves the cellulose and hemicellulose fibers in the cell wall [76]. The characterization of the concentration of these compounds directly influences the contents of fermentable sugars, fundamental components in the fermentation processes for the synthesis and properties of biodiesel.

Furthermore, constituents such as carbon, hydrogen, oxygen, nitrogen, chlorine, and sulfur are verified in high concentrations in oil palm biomass [75]. Studies showed that oil palm kernel contains large amounts of phenolic compounds, predominantly p-hydroxybenzoic and protocatechuic acids [81]. This biomass has approximately 80% saturated fatty acids, of which lauric acid is the most abundant (48.53%) [82]. Moreover, the adoption of integrated technologies, such as pyrolysis and microwave procedures, results in a significant increase in phenolic compounds and, consequently, in higher bio-oil productivity [83]. The same scenario was verified for the application of the combination of pyrolysis, gas chromatography, and mass spectrometry in oil palm mesocarp fibers [84]. The integrative strategy resulted in the determination of a multiplicity of compounds, such as acids (acetic, lauric, etc.), methyl palmitate, phenols, etc. A study aiming at the optimization of succinic acid production based on oil palm trunk with *Actinobacillus succinogenes* 130Z developed a biochemical composition survey [85]. The study reported the concentration of a range of available compounds, predominantly nutrients, such as potassium (4444.85 μg/g, sodium (284.68 μg/g, magnesium (278.76 μg/g, etc.,amino acids, such as aspartic acid (2630 μg/g, lysine (2255 μg/g, arginine (1970 μg/g, etc.; and vitamins, such as vitamin B7, biotin (200 μg/g, vitamin A, retinol (87.31 μg/g, vitamin C, ascorbic acid (36.95 μg/g, etc.

Table 2 Lignocellulosic composition of palm biomasses

	Biomass	Cellulose	Hemicellulose	Lignin	Reference
Composition (wt.%)	Kernel shell	27.7	21.6	44.0	[77]
	Mesocarp fibers	29.9	33.5	21.6	[78]
	Trunk	39.3	30.3	30.3	[79]
	Frond	56.0	28.0	20.5	[80]

5 Biodiesel Production from Palm Biomass Wastes

Palm oil is a valuable resource widely employed in a range of fields of study and supply chain components. A massive portion of indispensable products for human consumption contains palm oil as a predominant element, such as hygiene products, essential cosmetics, and food-related products [75]. Nonetheless, eedstocks from a range of plant species have been applied as potential sources for biodiesel production in key countries in the global economy [86]. Accordingly, palm oil is a major vegetable oil producer in the world, representing approximately 35% of all oil produced, and is appointed as one of the main natural sources for the synthesis of biodiesel [87]. Aiming at enhancing the production system and optimizing the value addition of the species, the use of palm oil from different fragments of the plants, such as shells, trunks, and fronds, has been largely explored over the years, evidencing the species as one of the pioneers in the biodiesel production chain [88].

According to the frond biomasses, a study conducted under the application of a fast pyrolysis procedure showed that oil yields of up to 36.4% mass were obtained under process conditions of increasing temperature up to 350 °C and thermal oil as a heat transporter [89]. The product obtained was characterized by a water content of 52.77% and a pH of 2.2, as well as high concentrations of carboxylic acid, acetic acid, and phenols. Moreover, a combination of saccharification and *Saccharomyces cerevisiae* TISTR5048 fermentation processes in a batch mode was explored in pretreated (sodium hydroxide, sulfuric acid, and sodium hydroxide in hydrogen peroxide) palm frond hydrolysates [90]. At high residual cellulose contents, sugar concentrations of up to 56.9 g/L were verified, obtaining approximately 17.5 g/L of ethanol. Finally, frond biomass under enzymatic hydrolysis with combinations of cellulase and hemicellulose enzymes and cellulase or hemicellulase applied individually, in process conditions of 50 ºC for 150 min, was used for ethanol synthesis [91]. The fermentation (*Saccharomyces cerevisiae* HC 10) of materials pretreated with cellulase and hemicellulose enzymes associated reported ethanol yields of 79.4% and 7.23 g/L after 24 h of the fermentation procedure.

Furthermore, the oil palm trunk is widely explored as a relevant resource for the biorefinery-based industry. An investigative study showed that biomass is an effective raw material for the synthesis of enzymes, ethanol, and acetic acid [92]. According to the ethanol production, biomass hydrolyzed under incubation at 50 °C for 15 h by crude enzymes from a range of solid-state fermented fungal species resulted in a maximum yield of 4.1 g/L. Besides, Indonesian scientific research reported that the efficient ethanol conversion process (EECP) resulted in yields of 22.3 g ethanol and maximum conversion of up to 93.5% in a combined operation with enzymatic hydrolysis procedures (50 °C and 200 rpm for 24 h) and ethanol fermentation with *Saccharomyces cerevisiae* (24858 ATCC) before the hydrothermal treatment (deionized water at 160–200 °C in intervals of 10 °C for 30 min) [93]. The methodological procedure adopted resulted in the potential of biodiesel production with cost reduction and process viability.

Additionally, oil palm mesocarp fiber wastes are also a valuable feedstock once the high contents of lignocellulose are considered in its constitution [94]. The dominant contents of oil, lignocellulose, acids, and monosaccharides contribute to adding value to these by-products and improving biodiesel production [75]. One of the main operations applied to agricultural residues to obtain bioethanol is the application of pressurized fluids. A recent study evidenced that the application of pressurized fluid technology in oil palm fiber with water as a process solvent under conditions of 210 °C, reaction times of 15 min at 60 bar, and solvent flow rate of 0.75 mL/min was sufficient for the generation of 30.47 mg/mL of xylose [95].

Finally, Ejeromedoghene [96] investigated the biodiesel characterization and yields from palm kernel oil (initial acid value of 8.4 mg $KOH{\cdot}g^{-1}$) via acid transesterification. Under 60 °C, methanol to oil molar ratio of 15:1, 15 vol% of concentrated acetic acid, a yield of 98% was accomplished after 30 min of reaction. The biodiesel synthesized was characterized by the author according to ASTM standards, obtaining an interesting reduction in the acid value (0.64 $KOH{\cdot}g^{-1}$) and on-spec values for specific gravity (0.889 $kg{\cdot}L^{-1}$), moisture (not detected), flashpoint (136 °C), and pour point (9 °C). The results presented demonstrated that the technique and raw material examined have the potential to be used in scales superior to lab one. In similar research, Kareem et al. [97] examined the preparation of biodiesel from palm kernel oil catalyzed by free lipases, obtained from the microorganism *Aspergillus niger* by solid-state fermentation. Using a statistical design, the authors evaluated the reaction parameters "process temperature" (30 to 40 °C), "methanol to oil molar ratio" (1:1 to 3:1), and "enzyme load" (2.5 to 7.5 wt.%), keeping the biodiesel (FAME) yield as the response variable. Maximum FAME yields (>90% after 48 h of reaction) were observed at 40 °C, methanol to oil molar ratio of 3:1 and 7.5 wt.% of lipase added to the system. In enzyme-catalyzed systems for the production of biodiesel, the parameters "reaction temperature" and "methanol to oil molar ratio" are crucial points to be analyzed in order to obtain satisfactory biodiesel yields, since high values of these variables can impact the activity of the biocatalyst and consequently harm the yield of the process [38].

Farabi et al. [98] synthesized a solid catalyst based on a sulfonated carbon compound using palm kernel shell to be employed in an esterification reaction of palm fatty acid distillate to biodiesel preparation. The authors used a process of sulfonation with a carbon source (70 °C for 4 h) to attach the sulfonate group ($-SO_3H$) group on the catalyst surface. The authors also verified that the sulfonated catalyst of palm kernel shell presented high acidity (14.4 $mmol{\cdot}g^{-1}$), which indicates the compound has high catalytic activity. The catalytic capacity of the synthesized material was tested in an esterification reaction for biodiesel production: at 65 °C, catalyst concentration of 4 wt.%, methanol to feedstock molar ratio of 15:1 and 1 h of reaction, 95% of FAME yield was obtained, with a free fatty acid conversion of 97%. Moreover, the catalyst could be reused up to 4 times, demonstrating the potential of the compound in reducing the production costs of biodiesel synthesis.

6 Conclusions and Future Outlooks

The current scenario of biodiesel production globally reports the importance of natural sources as predominant resources in the planet's energy fraction. The feasibility of using palm oil as a potential agent in the bioenergy sector is in line with the basic principles that sustain sustainability and highlight the serious threat of environmental imbalance on the planet. The impacts of a troubled relationship between humanity and the resources available on the planet have caused a marked degradation and an imbalance in the efficient functioning of countless macro and microprocesses. This chapter reinforced palm oil and its by-products as potential resources for biodiesel development, highlighting the invaluable importance of palm oil and palm oil-based industries to prospect a sustainable promise to encourage discussions and studies for the power generation approach. Furthermore, the expansion of exploration of techniques and technologies contributes to the proposed theme being considered in the scientific and industrial fields and provides a formidable challenge to maximize the potential of environmentally friendly strategies in the coming years. The increase in palm cultivation over the years shows not only the potential for oil production but of lignocellulose-based waste generation. Palm oil responds to over 40% of the total oil produced globally. This panorama is evidenced by the proportion of the increase in oil and the cultivation area in a large area worldwide. In addition, oil productivity per plant is significantly higher than that seen in widely traded commodities.

Finally, there are still many challenges in the bioenergy sector and terms of resilience in the direction of the industry toward the exploitation of bio-friendly resources. Even with broad academic and industrial recognition, the use of feedstocks as precursor agents for biodiesel production needs continuous discussion and investment. Perspectives about the palm oil approach in terms of economic and sustainable viability must be investigated to accomplish higher improvement and effectiveness in the use of this product and its bio-waste for power generation in the energy and biofuels approach. Accordingly, essential notes are highlighted as facilitating the understanding of future perspectives and challenges in the field of palm oil and palm oil-based biodiesel:

- Constant investments in innovative technologies;
- Incentive for the cultivation of palm to producers and agribusiness-based companies;
- Specification and operation of social and organizational policies and programs aimed at enhancing the biodiesel production;
- Optimization of industrial process steps in the biodiesel production background;
- Use of catalysts as process improvement agents;
- Improvement of product logistics, transport, and distribution structures.

References

1. Mathew GM, Raina D, Narisetty V et al (2021) Recent advances in biodiesel production: challenges and solutions. Sci Total Environ 794:148751. https://doi.org/10.1016/j.scitotenv.2021.148751
2. Mobin Siddique MB, Khairuddin N, Ali NA et al (2021) A comprehensive review on the application of bioethanol/biodiesel in direct injection engines and consequential environmental impact. Clean Eng Technol 3:100092. https://doi.org/10.1016/j.clet.2021.100092
3. Knothe G, Razon LF (2017) Biodiesel fuels. Prog Energy Combust Sci 58:36–59. https://doi.org/10.1016/j.pecs.2016.08.001
4. Wancura JHC, Fantinel AL, Ugalde GA et al (2021) Semi-continuous production of biodiesel on pilot scale via enzymatic hydroesterification of waste material: process and economics considerations. J Clean Prod 285.https://doi.org/10.1016/j.jclepro.2020.124838
5. Qadeer MU, Ayoub M, Komiyama M et al (2021) Review of biodiesel synthesis technologies, current trends, yield influencing factors and economical analysis of supercritical process. J Clean Prod 309:127388. https://doi.org/10.1016/j.jclepro.2021.127388
6. Wancura JHC, Tres MV, Jahn SL, Oliveira JV (2020) Lipases in liquid formulation for biodiesel production: current status and challenges. Biotechnol Appl Biochem 67:648–667. https://doi.org/10.1002/bab.1835
7. Bin MMN, Tan YH, Seow YX et al (2021) Evaluation on feedstock, technologies, catalyst and reactor for sustainable biodiesel production: a review. J Ind Eng Chem 98:60–81. https://doi.org/10.1016/j.jiec.2021.03.036
8. Malik S, Shahid A, Liu C-G et al (2021) Developing fourth-generation biofuels secreting microbial cell factories for enhanced productivity and efficient product recovery: a review. Fuel 298:120858. https://doi.org/10.1016/j.fuel.2021.120858
9. Dey S, Reang NM, Das PK, Deb M (2021) A comprehensive study on prospects of economy, environment, and efficiency of palm oil biodiesel as a renewable fuel. J Clean Prod 286:124981. https://doi.org/10.1016/j.jclepro.2020.124981
10. Castanheira ÉG, Freire F (2017) Environmental life cycle assessment of biodiesel produced with palm oil from Colombia. Int J Life Cycle Assess 22:587–600. https://doi.org/10.1007/s11367-016-1097-6
11. Ong ES, Rabbani AH, Habashy MM et al (2021) Palm oil industrial wastes as a promising feedstock for biohydrogen production: a comprehensive review. Environ Pollut 291:118160. https://doi.org/10.1016/j.envpol.2021.118160
12. Ahmad FB, Zhang Z, Doherty WOS, O'Hara IM (2019) The prospect of microbial oil production and applications from oil palm biomass. Biochem Eng J 143:9–23. https://doi.org/10.1016/j.bej.2018.12.003
13. Budžaki S, Miljić G, Sundaram S et al (2018) Cost analysis of enzymatic biodiesel production in small-scaled packed-bed reactors. Appl Energy 210:268–278. https://doi.org/10.1016/j.apenergy.2017.11.026
14. Allami HA, Nayebzadeh H (2021) The assessment of the engine performance and emissions of a diesel engine fueled by biodiesel produced using different types of catalyst. Fuel 305:121525. https://doi.org/10.1016/j.fuel.2021.121525
15. Intasit R, Cheirsilp B, Louhasakul Y et al (2020) Valorization of palm biomass wastes for biodiesel feedstock and clean solid biofuel through non-sterile repeated solid-state fermentation. Bioresour Technol 298:122551. https://doi.org/10.1016/j.biortech.2019.122551
16. Intasit R, Cheirsilp B, Louhasakul Y, Boonsawang P (2020b) Consolidated bioprocesses for efficient bioconversion of palm biomass wastes into biodiesel feedstocks by oleaginous fungi and yeasts. Bioresour Technol 315.https://doi.org/10.1016/j.biortech.2020.123893
17. Quayson E, Amoah J, Rachmadona N et al (2020) Valorization of palm biomass waste into carbon matrices for the immobilization of recombinant *Fusarium heterosporum* lipase towards palm biodiesel synthesis. Biomass Bioenergy 142. https://doi.org/10.1016/j.biombioe.2020.105768

18. Tan C, Saritpongteeraka K, Kungsanant S et al (2018) Low temperature hydrothermal treatment of palm fiber fuel for simultaneous potassium removal, enhanced oil recovery and biogas production. Fuel 234:1055–1063. https://doi.org/10.1016/j.fuel.2018.07.137
19. Maneerung T, Kawi S, Dai Y, Wang CH (2016) Sustainable biodiesel production via transesterification of waste cooking oil by using CaO catalysts prepared from chicken manure. Energy Convers Manag 123:487–497. https://doi.org/10.1016/j.enconman.2016.06.071
20. Krishania N, Rajak U, Nath Verma T et al (2020) Effect of microalgae, tyre pyrolysis oil and *Jatropha* biodiesel enriched with diesel fuel on performance and emission characteristics of CI engine. Fuel 278:118252. https://doi.org/10.1016/j.fuel.2020.118252
21. Laesecke J, Ellis N, Kirchen P (2017) Production, analysis and combustion characterization of biomass fast pyrolysis oil—Biodiesel blends for use in diesel engines. Fuel 199:346–357. https://doi.org/10.1016/j.fuel.2017.01.093
22. Shiroudi A, Hirao K, Yoshizawa K et al (2020) A computational study on the kinetics of pyrolysis of isopropyl propionate as a biodiesel model: DFT and ab initio investigation. Fuel 281:118798. https://doi.org/10.1016/j.fuel.2020.118798
23. Abdelfattah MSH, Abu-Elyazeed OSM, Abd El Mawla E, Abdelazeem MA (2018) On biodiesels from castor raw oil using catalytic pyrolysis. Energy 143:950–960. https://doi.org/10.1016/j.energy.2017.09.095
24. Amin A (2019) Review of diesel production from renewable resources: catalysis, process kinetics and technologies. Ain Shams Eng J 10:821–839. https://doi.org/10.1016/J.ASEJ.2019.08.001
25. Sun Y, Li C, Zhang S et al (2021) Pyrolysis of soybean residue: understanding characteristics of the products. Renew Energy 174:487–500. https://doi.org/10.1016/J.RENENE.2021.04.063
26. Pandit PR, Fulekar MH (2019) Biodiesel production from microalgal biomass using CaO catalyst synthesized from natural waste material. Renew Energy 136:837–845. https://doi.org/10.1016/j.renene.2019.01.047
27. Ferreira MFP, Oliveira BFH, Pinheiro WBS et al (2020) Generation of biofuels by slow pyrolysis of palm empty fruit bunches: optimization of process variables and characterization of physical-chemical products. Biomass Bioenerg 140:105707. https://doi.org/10.1016/J.BIOMBIOE.2020.105707
28. Lee XJ, Lee LY, Hiew BYZ et al (2020) Valorisation of oil palm wastes into high yield and energy content biochars via slow pyrolysis: multivariate process optimisation and combustion kinetic studies. Mater Sci Energy Technol 3:601–610. https://doi.org/10.1016/J.MSET.2020.06.006
29. Demaman Oro CE, Bonato M, Oliveira JV et al (2019) A new approach for salts removal from crude glycerin coming from industrial biodiesel production unit. J Environ Chem Eng 7:102883. https://doi.org/10.1016/j.jece.2019.102883
30. Battiston CSZ, Ficanha AMM, Oro CED et al (2021) In situ calb enzyme immobilization in mesoporous material type MCM-48 synthesis using ionic solid [C14MI]Cl as structure-directing agent. Appl Biochem Biotechnol.https://doi.org/10.1007/s12010-021-03648-z
31. Ficanha AMM, Antunes A, Oro CED et al (2020) Immobilization of *Candida antarctica* B (CALB) in silica aerogel: morphological characteristics and stability. Biointerface Res Appl Chem 10:6744–6756. https://doi.org/10.33263/BRIAC106.67446756
32. Tan X, Zhang H, Li H, Yang S (2022) Electrovalent bifunctional acid enables heterogeneously catalytic production of biodiesel by (trans)esterification of non-edible oils. Fuel 310:122273. https://doi.org/10.1016/J.FUEL.2021.122273
33. Widayat, Wibowo ADK, Hadiyanto (2013) Study on production process of biodiesel from rubber seed (*Hevea brasiliensis*) by in situ (trans)esterification method with acid catalyst. Energy Procedia 32:64–73.https://doi.org/10.1016/j.egypro.2013.05.009
34. Wancura JHC, Rosset DV, Mazutti MA et al (2019) Improving the soluble lipase–catalyzed biodiesel production through a two-step hydroesterification reaction system. Appl Microbiol Biotechnol 103:7805–7817. https://doi.org/10.1007/s00253-019-10075-y
35. Ficanha AMM, Oro CED, Franceschi E et al (2021) Evaluation of different ionic liquids as additives in the immobilization of lipase CAL B by Sol-Gel technique. Appl Biochem Biotechnol 193:2162–2181. https://doi.org/10.1007/s12010-021-03533-9

36. Bordin I, de Aguiar PV, Demaman Oro CE et al (2021) Nb-MCM-type mesoporous material synthesis using ionic solid as structure-directing agent for in situ lipase immobilization. Appl Biochem Biotechnol. https://doi.org/10.1007/s12010-020-03484-7
37. Ficanha AMM, Antunes A, Oro CED et al (2021a) Immobilization of lipase CALB in organically modified silica. Biointerface Res Appl Chem 11:7814–7825. https://doi.org/10.33263/BRIAC111.78147825
38. Wancura JHC, Rosset DV, Tres MV et al (2018) Production of biodiesel catalyzed by lipase from *Thermomyces lanuginosus* in its soluble form. Can J Chem Eng 96:2361–2368. https://doi.org/10.1002/cjce.23146
39. Aboelazayem O, Gadalla M, Alhajri I, Saha B (2021) Advanced process integration for supercritical production of biodiesel: residual waste heat recovery via organic Rankine cycle (ORC). Renew Energy 164:433–443. https://doi.org/10.1016/j.renene.2020.09.058
40. Andreo-Martínez P, Ortiz-Martínez VM, García-Martínez N et al (2020) Production of biodiesel under supercritical conditions: state of the art and bibliometric analysis. Appl Energy 264:114753. https://doi.org/10.1016/j.apenergy.2020.114753
41. Jafari A, Esmaeilzadeh F, Mowla D et al (2021) New insights to direct conversion of wet microalgae impregnated with ethanol to biodiesel exploiting extraction with supercritical carbon dioxide. Fuel 285:119199. https://doi.org/10.1016/j.fuel.2020.119199
42. Ridwan I, Katsuragawa N, Tamura K (2020) Process optimization, reaction kinetics, and thermodynamics studies of water addition on supercritical methyl acetate for continuous biodiesel production. J Supercrit Fluids 166:105038. https://doi.org/10.1016/j.supflu.2020.105038
43. Roschat W, Phewphong S, Khunchalee J, Moonsin P (2018) Biodiesel production by ethanolysis of palm oil using SrO as a basic heterogeneous catalyst. Mater Today Proc 5:13916–13921. https://doi.org/10.1016/j.matpr.2018.02.040
44. Maneechakr P, Karnjanakom S (2021) Systematic production of biodiesel fuel from palm oil over porous K_2O-CaO catalyst derived from waste chicken eggshell via RSM/kinetic/thermodynamic studies. J Environ Chem Eng 9:106542. https://doi.org/10.1016/j.jece.2021.106542
45. Zhang P, Chen X, Leng Y et al (2020) Biodiesel production from palm oil and methanol via zeolite derived catalyst as a phase boundary catalyst: an optimization study by using response surface methodology. Fuel 272:117680. https://doi.org/10.1016/j.fuel.2020.117680
46. Qu T, Niu S, Zhang X et al (2021) Preparation of calcium modified Zn-Ce/Al_2O_3 heterogeneous catalyst for biodiesel production through transesterification of palm oil with methanol optimized by response surface methodology. Fuel 284:118986. https://doi.org/10.1016/j.fuel.2020.118986
47. Zulqarnain YMHM, Ayoub M et al (2021) Solvent extraction and performance analysis of residual palm oil for biodiesel production: experimental and simulation study. J Environ Chem Eng 9:105519. https://doi.org/10.1016/J.JECE.2021.105519
48. Thushari I, Babel S, Samart C (2019) Biodiesel production in an autoclave reactor using waste palm oil and coconut coir husk derived catalyst. Renew Energy 134:125–134. https://doi.org/10.1016/j.renene.2018.11.030
49. Nongbe MC, Ekou T, Ekou L et al (2017) Biodiesel production from palm oil using sulfonated graphene catalyst. Renew Energy 106:135–141. https://doi.org/10.1016/j.renene.2017.01.024
50. Basyouny MG, Abukhadra MR, Alkhaledi K et al (2021) Insight into the catalytic transformation of the waste products of some edible oils (corn oil and palm oil) into biodiesel using MgO/clinoptilolite green nanocomposite. Mol Catal 500:111340. https://doi.org/10.1016/J.MCAT.2020.111340
51. Loh JM, Amelia GW et al (2021) Improved biodiesel production from sludge palm oil catalyzed by a low-cost liquid lipase under low-input process conditions. Renew Energy 177:348–358. https://doi.org/10.1016/j.renene.2021.05.138
52. Matinja AI, Mohd Zain NA, Suhaimi MS, Alhassan AJ (2019) Optimization of biodiesel production from palm oil mill effluent using lipase immobilized in PVA-alginate-sulfate beads. Renew Energy 135:1178–1185. https://doi.org/10.1016/j.renene.2018.12.079

53. Shahedi M, Habibi Z, Yousefi M et al (2021) Improvement of biodiesel production from palm oil by co-immobilization of *Thermomyces lanuginosa* lipase and *Candida antarctica* lipase B: optimization using response surface methodology. Int J Biol Macromol 170:490–502. https://doi.org/10.1016/j.ijbiomac.2020.12.181
54. Guo J, Sun S, Liu J (2020) Conversion of waste frying palm oil into biodiesel using free lipase A from *Candida antarctica* as a novel catalyst. Fuel 267:117323. https://doi.org/10.1016/j.fuel.2020.117323
55. Akkarawatkhoosith N, Kaewchada A, Jaree A (2019) Simultaneous development of biodiesel synthesis and fuel quality via continuous supercritical process with reactive co-solvent. Fuel 237:117–125. https://doi.org/10.1016/j.fuel.2018.09.077
56. Ndiaye M, Arhaliass A, Legrand J et al (2020) Reuse of waste animal fat in biodiesel: Biorefining heavily-degraded contaminant-rich waste animal fat and formulation as diesel fuel additive. Renew Energy 145:1073–1079. https://doi.org/10.1016/j.renene.2019.06.030
57. Toldra-Reig F, Mora L, Toldra F (2020) Applied sciences trends in biodiesel production from animal fat waste. Appl Sci 10:1–17
58. Purandaradas A, Silambarasan T, Murugan K et al (2018) Development and quantification of biodiesel production from chicken feather meal as a cost-effective feedstock by using green technology. Biochem Biophys Reports 14:133–139. https://doi.org/10.1016/j.bbrep.2018.04.012
59. Lin C, Tsai S (2015) Production of biodiesel from chicken wastes by various. J Energy Southern Africa 26:36–45
60. Soares Dias AP, Ramos M, Catarino M, Costa Pereira MF (2020) Biodiesel by co-processing animal fat/vegetable oil mixtures over basic heterogeneous Ca catalyst. Clean Eng Technol 1:1–6. https://doi.org/10.1016/j.clet.2020.100012
61. Gad MS, EL-Seesy AI, Abu Hashish HM et al (2021) Combustion and emissions aspects of a diesel engine working with sheep fat oil biodiesel-diesel blends. Case Stud Therm Eng 26:101162. https://doi.org/10.1016/j.csite.2021.101162
62. Hájek M, Vávra A, de Paz CH, Kocík J (2021) The catalysed transformation of vegetable oils or animal fats to biofuels and bio-lubricants: a review. Catalysts 11:1–30. https://doi.org/10.3390/catal11091118
63. Hönig V, Strouhal J, Linhart Z (2016) Biodiesel from waste animal fat: efficient fuel of the future? Int Adv Econ Res 22:465–466. https://doi.org/10.1007/s11294-016-9603-y
64. Ramos M, Dias APS, Puna JF et al (2019) Biodiesel production processes and sustainable raw materials. Energies 12.https://doi.org/10.3390/en12234408
65. Vargas-Ibáñez LT, Cano-Gómez JJ, Zwolinski P, Evrard D (2020) Environmental assessment of an animal fat based biodiesel: defining goal, scope and life cycle inventory. Procedia CIRP 90:215–219. https://doi.org/10.1016/j.procir.2020.02.053
66. Yaqoob H, Teoh YH, Sher F et al (2021) Potential of waste cooking oil biodiesel as renewable fuel in combustion engines: a review. Energies 14. https://doi.org/10.3390/en14092565
67. Wang XL, Zhou JJ, Sun YQ, Xiu ZL (2019) Bioconversion of raw glycerol from waste cooking-oil-based biodiesel production to 1,3-propanediol and lactate by a microbial consortium. Front Bioeng Biotechnol 7:1–13. https://doi.org/10.3389/fbioe.2019.00014
68. Degfie TA, Mamo TT, Mekonnen YS (2019) Optimized biodiesel production from waste cooking oil (WCO) using calcium oxide (CaO) nano-catalyst. Sci Rep 9:1–8. https://doi.org/10.1038/s41598-019-55403-4
69. Sahar SS, Iqbal J et al (2018) Biodiesel production from waste cooking oil: an efficient technique to convert waste into biodiesel. Sustain Cities Soc 41:220–226. https://doi.org/10.1016/j.scs.2018.05.037
70. Khan N, Park SH, Kadima L et al (2021) Locally sustainable biodiesel production from waste cooking oil and grease using a deep eutectic solvent: characterization, thermal properties, and blend performance. ACS Omega 6:9204–9212. https://doi.org/10.1021/acsomega.1c00556
71. Borah MJ, Das A, Das V et al (2019) Transesterification of waste cooking oil for biodiesel production catalyzed by Zn substituted waste egg shell derived CaO nanocatalyst. Fuel 242:345–354. https://doi.org/10.1016/j.fuel.2019.01.060

72. Ali CH, Asif AH, Iqbal T et al (2018) Improved transesterification of waste cooking oil into biodiesel using calcined goat bone as a catalyst. Energy Sources, Part A Recover Util Environ Eff 40:1076–1083. https://doi.org/10.1080/15567036.2018.1469691
73. Erchamo YS, Mamo TT, Workneh GA, Mekonnen YS (2021) Improved biodiesel production from waste cooking oil with mixed methanol–ethanol using enhanced eggshell-derived CaO nano-catalyst. Sci Rep 11:1–12. https://doi.org/10.1038/s41598-021-86062-z
74. Nadeem F, Bhatti IA, Ashar A et al (2021) Eco-benign biodiesel production from waste cooking oil using eggshell derived MM-CaO catalyst and condition optimization using RSM approach. Arab J Chem 14:103263. https://doi.org/10.1016/j.arabjc.2021.103263
75. Kaniapan S, Hassan S, Ya H et al (2021) The utilisation of palm oil and oil palm residues and the related challenges as a sustainable alternative in biofuel, bioenergy, and transportation sector: a review. Sustain 13. https://doi.org/10.3390/su13063110
76. Rizal NFAA, Ibrahim MF, Zakaria MR et al (2018) Pre-treatment of oil palm biomass for fermentable sugars production. Molecules 23:1–14. https://doi.org/10.3390/molecules23061381
77. Zainal NH, Aziz AA, Idris J et al (2017) Microwave-assisted pre-carbonisation of palm kernel shell produced charcoal with high heating value and low gaseous emission. J Clean Prod 142:2945–2949. https://doi.org/10.1016/j.jclepro.2016.10.176
78. Saidu M, Yuzir A, Salim MR et al (2014) Biological pre-treated oil palm mesocarp fibre with cattle manure for biogas production by anaerobic digestion during acclimatization phase. Int Biodeterior Biodegrad 95:189–194. https://doi.org/10.1016/j.ibiod.2014.06.014
79. Nasser RA, Salem MZM, Hiziroglu S et al (2016) Chemical analysis of different parts of date palm (*Phoenix dactylifera* L.) using ultimate, proximate and thermo-gravimetric techniques for energy production. Energies 9. https://doi.org/10.3390/en9050374
80. Mohamad Ikubar MR, Abdul Manan M, Md. Salleh M, Yahya A (2018) Solid-state fermentation of oil palm frond petiole for lignin peroxidase and xylanase-rich cocktail production. 3 Biotech 8:1–11. https://doi.org/10.1007/s13205-018-1268-1
81. Tsouko E, Alexandri M, Fernandes KV et al (2019) Extraction of phenolic compounds from palm oil processing residues and their application as antioxidants. Food Technol Biotechnol 57:29–38. https://doi.org/10.17113/ftb.57.01.19.5784
82. Egbuna SO, Nwachukwu UJ, Agu CM et al (2021) Production of biolubricant samples from palm kernel oil using different chemical modification approaches. Eng Reports 3:1–19. https://doi.org/10.1002/eng2.12422
83. Ogunkanmi JO, Kulla DM, Omisanya NO et al (2018) Extraction of bio-oil during pyrolysis of locally sourced palm kernel shells: effect of process parameters. Case Stud Therm Eng 12:711–716. https://doi.org/10.1016/j.csite.2018.09.003
84. Babinszki B, Jakab E, Terjék V et al (2021) Thermal decomposition of biomass wastes derived from palm oil production. J Anal Appl Pyrolysis 155. https://doi.org/10.1016/j.jaap.2021.105069
85. Bukhari NA, Loh SK, Nasrin AB et al (2019) Compatibility of utilising nitrogen-rich oil palm trunk sap for succinic acid fermentation by *Actinobacillus succinogenes* 130Z. Bioresour Technol 293.https://doi.org/10.1016/j.biortech.2019.122085
86. Mahlia TMI, Ismail N, Hossain N et al (2019) Palm oil and its wastes as bioenergy sources: a comprehensive review. Environ Sci Pollut Res. https://doi.org/10.1007/s11356-019-04563-x
87. Meijaard E, Sheil D (2019) The moral minefield of ethical oil palm and sustainable development. Front Glob Chang 2. https://doi.org/10.3389/ffgc.2019.00022
88. Kurnia JC, Jangam S V., Akhtar S et al (2016) Advances in biofuel production from oil palm and palm oil processing wastes: a review. Biofuel Res J 3:332–346. https://doi.org/10.18331/BRJ2016.3.1.3
89. Solikhah MD, Raksodewanto AA, Kismanto A et al (2017) Development of bio-fuel from palm frond via fast pyrolysis. IOP Conf Ser Earth Environ Sci 65. https://doi.org/10.1088/1755-1315/65/1/012014
90. Kumneadklang S, Larpkiattaworn S, Niyasom C, O-Thong S (2015) Bioethanol Production from Oil Palm Frond by Simultaneous Saccharification and Fermentation. Elsevier B.V.

91. Farah Amani AH, Toh SM, Tan JS, Lee CK (2018) The efficiency of using oil palm frond hydrolysate from enzymatic hydrolysis in bioethanol production. Waste and Biomass Valorization 9:539–548. https://doi.org/10.1007/s12649-017-0005-z
92. Nutongkaew T, Prasertsan P, Leamdum C et al (2020) Bioconversion of oil palm trunk residues hydrolyzed by enzymes from newly isolated fungi and use for ethanol and acetic acid production under two-stage and simultaneous fermentation. Waste Biomass Valor 11:1333–1347. https://doi.org/10.1007/s12649-019-00678-x
93. Eom IY, Yu JH, Jung CD, Hong KS (2015) Efficient ethanol production from dried oil palm trunk treated by hydrothermolysis and subsequent enzymatic hydrolysis. Biotechnol Biofuels 8:1–11. https://doi.org/10.1186/s13068-015-0263-6
94. Nordin NIAA, Ariffin H, Andou Y et al (2013) Modification of oil palm mesocarp fiber characteristics using superheated steam treatment. Molecules 18:9132–9146. https://doi.org/10.3390/molecules18089132
95. De Oliveira ALPC, Da Antônio Clinton CG, Almeida PS et al (2018) The use of compressed fluids to obtain biocomposites from palm oil fiber (*Elaeis* sp.). Brazilian J Chem Eng 35:353–361. https://doi.org/10.1590/0104-6632.20180352s20160374
96. Ejeromedoghene O (2021) Acid-catalyzed transesterification of palm kernel oil (PKO) to biodiesel. Mater Today Proc 47:1580–1583. https://doi.org/10.1016/j.matpr.2021.04.042
97. Kareem SO, Falokun EI, Balogun SA et al (2017) Enzymatic biodiesel production from palm oil and palm kernel oil using free lipase. Egypt J Pet 26:635–642. https://doi.org/10.1016/j.ejpe.2016.09.002
98. Farabi MSA, Ibrahim ML, Rashid U, Taufiq-Yap YH (2019) Esterification of palm fatty acid distillate using sulfonated carbon-based catalyst derived from palm kernel shell and bamboo. Energy Convers Manag 181:562–570. https://doi.org/10.1016/j.enconman.2018.12.033

Hydrogen as an Alternative Biofuel Through Gasification Process: Comparative Study of the EU and Turkey

F. M. Alptekin and M. S. Çeliktaş

Abstract The global energy demand has approximately reduplicated since 1980 and the transport sector occupies a considerable amount of this demand. In addition to the transportation sector, fossil-based fuels used in industry and buildings adversely affect the environment and cause global warming due to the CO_2 emissions they release. Hence, it is important to use alternative fuels instead of these fossil fuels. Hydrogen is considered one of the future alternative fuels due to its unique properties. In this work, waste/biomass to hydrogen as an alternative biofuel is evaluated. The study aims to examine waste/biomass as renewable resources for producing hydrogen. Brief information on waste/biomass resources and hydrogen energy conversion technologies are explained. A comparative case study for hydrogen production via the gasification process is examined for the EU countries and Turkey. The findings are assessed and suggestions about the implementation of hydrogen production via the gasification process are presented.

Keywords Alternative fuel · Hydrogen · Gasification · EU · Turkey

1 Introduction

To supply energy demand and energy security and decrease the negative environmental effect of fossil resources, sustainable and scalable energy sources are required [1, 2]. The transportation sector is one of the major consumption areas of energy. According to the International Energy Agency (IEA), the transportation sector uses 20% of global energy at present [3]. The transportation sector is a greater contributor to emitting CO_2 with sharing three-quarters of total emissions. Hence, decarbonizing the transport sector is essential [4]. To overcome the detrimental impact of oil and fossil fuels, developing and adopting a new policy regarding finding alternative fuels is necessary [5, 6].

F. M. Alptekin · M. S. Çeliktaş (✉)
Ege University Solar Energy Institute, 35100 İzmir, Turkey
e-mail: soner.celiktass@ege.edu.tr

A. Khan et al. (eds.), *Waste to Biofuel Technology*, Sustainable Materials and Technology, https://doi.org/10.1007/978-981-97-4561-6_2

Increasing greenhouse emissions and environmental awareness have contributed to the sustainable evaluation of waste, biomass, and different kinds of residues. Disposal, storage, and sorting were treatments for waste till the '90s while direct recycling of waste and residues was used after that date. Nevertheless, including different pollutants like heavy metals, volatile organic compounds, PAH, limited direct recycling of waste and residues. Being safer, reusing waste and biomass have gained importance for valorization them to energy, fuel, and other valuable materials [7].

Based on their physical form, wastes can be categorized as solid, liquid, and gaseous. Municipal solid waste, bio-medical waste, industrial waste [8], agricultural waste [9], hazardous waste [10], and e-waste are representative wastes of solid waste. Liquid waste includes sewage, refinery waste, wasted liquid byproducts, and liquid effluents. Gases from a stack emitted gaseous products, and foam are the type of gaseous wastes [11].

Agricultural and forestry residues, animal waste, municipal solid wastes, energy crops, and algae are sources of biomass [12]. Any other classification of biomass can be identified based on its nature as non-lignocellulosic or lignocellulosic. Two types can exist in different forms woody, herbaceous, aquatic debris, farming manure, and by-products, and other forms [13]. According to IEA, biomass-based technologies have been considered as "game changers" in the view of reducing consumption of fossil fuel and GHG emissions [12].

Although alternative fuels may vary depending on their origin and production process, their products consist of sustainable and clean production without excess CO_2 emissions [14]. Direct consumption of electric surplus and thermochemical conversion of feedstock is the main route to synthesize alternative fuels [15]. The first route is termed electro fuel which is carbon–neutral fuel and produced via various renewable energy sources electricity surplus as a form of gas, or liquid [16]. The second route is thermochemical conversion which produces alternative fuels and chemicals in gas or liquid form via converting waste materials or raw feedstock [17]. Among them, gaseous fuels such as syngas and biogas are the most outstanding alternative fuels for use in stationary plants while hydrogen is primarily featured as a gaseous fuel for application in mobile [14].

Among the aforementioned alternative fuels, hydrogen plays a critical role in decarbonizing energy systems [18]. Hydrogen is considered the energy carrier of the future due to having superior properties such as a higher heating value of 141.9 kJ/g (approximately 2.6 times greater than that of natural gas, 4.8 times greater than ethanol, and 3 times greater than that of gasoline) [19], more abundant in nature, non-toxic, non-create any emission, etc. [20–22].

Hydrogen can be produced via different technologies from both fossil-based energy sources and renewable energy sources such as wind, solar, biomass, electrolysis, etc. Gasification of coal or biomass is one of those technologies that includes thermochemical conversion of feedstock at high temperatures via a gasifying agent [23]. Hydrogen is categorized as "grey, blue and green hydrogen" based on its production methods [24]. In contrast to producing hydrogen via different technologies, biomass gasification has lower efficiency (39–48%) [25], but it is a good alternative to fossil-based sources due to being carbon-zero nature of biomass and green hydrogen production technology.

To overcome the unfavorable effects of fuel, countries all over the world have investigated new alternative fuels instead of using fossil-based fuels. The European Parliament and the Council established Directive 2014/94/EU for the deployment of alternative fuels infrastructure (AFID) [26]. Based on the AFID, the enhancement of an alternative fuels infrastructure in Europe that involves electricity for electric vehicles, hydrogen for fuel cells, and LNG for shipping was aimed [27–29]). On the other hand, Turkey has determined its target concerning energy under the Vision 2023 energy agenda to reduce its energy dependence and supply energy demand.

In this work, waste/biomass to hydrogen as an alternative biofuel is evaluated. The study aims to examine waste/biomass as renewable resources for producing hydrogen. After brief information on waste/biomass resources and hydrogen energy, in the following sections, conversion technologies are explained. A comparative case study for hydrogen production via the gasification process is examined for EU countries and Turkey. The findings are assessed and suggestions about the implementation of hydrogen production via the gasification process are presented.

2 Waste to Hydrogen Production Technologies

Physico-chemical, bio-chemical, and thermochemical are three main processes that are used for the conversion of biomass/waste to biofuel, energy, and valuable materials [30]. Quality and quantity of biomass feedstock, availability, cost, desired end products, and environmental concerns are factors that affect the selection of conversion technology [31]. In Fig. 1, types of potential biomass/waste feedstock are shown.

In the aspect of cost-effectiveness, agricultural residues, and municipal bio-waste have great potential. Also, energy crops candidate a good choice for producing hydrogen due to being non-edible crops and highest growth potential [33]. The high amount of carbohydrate and lipid content are important parameters that affect the determination of organic biomass for thermochemical conversion to achieve valuable products like H_2, bio-oil, and biochar [34]. In Fig. 2, conversion technologies of biomass/waste are shown.

In the view of producing hydrogen from biomass/waste, there are varied routes that are biochemical conversion, thermochemical conversion, and electrochemical. Nevertheless, electrochemical conversion technology is not feasible due to having a low conversion rate and requiring high energy input [11, 37]. In the following section, biochemical and thermochemical conversion technologies are explained.

Fig. 1 Type of potential biomass/waste feedstock. Revised from Liu et al. [32]

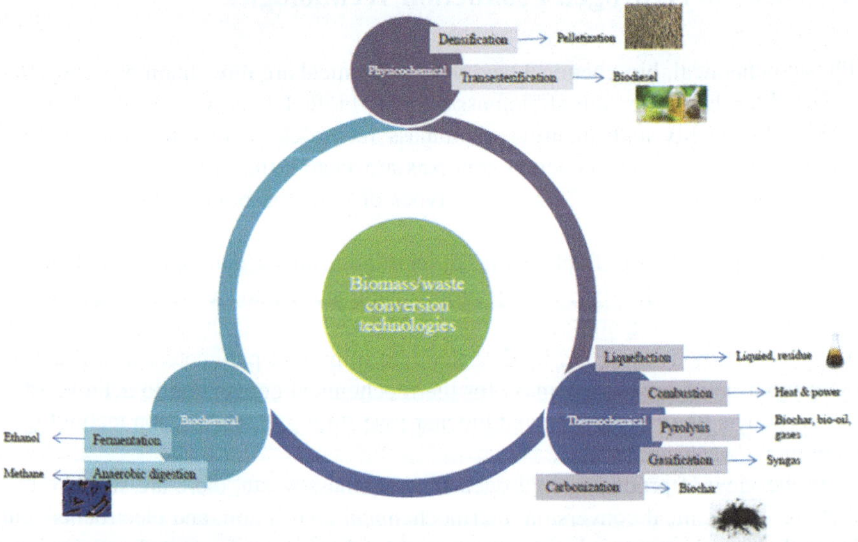

Fig. 2 Conversion technologies of biomass/waste to biofuel and energy are shown. Revised from [33, 35, 36]

2.1 *Biochemical Conversion*

Biological water gas shift reaction, dark fermentation, and photo-fermentation are three main routes that are applied in biochemical conversion. Enzymes are important for catalyzing H_2 production which affects each process. To synthesize H_2, microorganisms are necessary to produce enzymes [38]. Via microorganisms, bio-hydrogen can be produced from organic biomass [39]. The major biochemical process includes photolysis and fermentation which are divided into their sub-process. Direct and indirect bio-photolysis takes place in the bio-photolysis process while dark-, photo, and hybrid which is a combination of dark and photo take place in fermentation process [40]. In this section, the fermentation process is mentioned because of using biomass for hydrogen generation.

2.1.1 Fermentation

In dark fermentation, reducing sugar is obtained by hydrolyzing carbohydrates and then H_2, CO_2, fatty acids, and alcohols are achieved within anaerobic degradation of reducing sugar. Hydrogen-producing microbes such as *Clostridiaceae, Sporolactobacillaceae, Enterobacteriaceae, Bacillaceae, and Thermoanaerobacteriacea* are used for the process. Even if theoretically 1 mol of glucose generates 4 mol H_2, experimental results are generally lower than expected due to the formation of acetate, propionate, lactic acid, and alcohol [41].

The main drawback of the dark fermentation process is the consumption of hydrogen by Hydrogen-consuming bacteria (HCB). To increase the activity of Hydrogen-producing bacteria (HPB) and inhibit or eliminate HCBs, pretreatment methods are required [42].

Biohydrogen production by applying photo fermentation is an environmentally friendly way that suitable to conduct at ambient conditions. However, it has challenges regarding light conversion efficiency and lowers biohydrogen production [43]. In photo fermentation, organic compounds decompose in the presence of light supplied by photosynthetic bacteria or anaerobic bacteria strains, including *Rhodobium, Rhodobacter, Rhodospirillum,* and *Rhodopseudomonas,* biohydrogen is produced via nitrogenase-catalyzed reaction [44]. Lignocellulosic biomasses including energy crops, agricultural residues, forestry waste, forestry, industrial and domestic waste, algae, and any other animal manure are potential feedstock for producing biohydrogen through photo fermentation [43].

2.2 *Thermochemical Conversion*

In the aspect of utilizing any biomass and waste type, thermochemical conversion methods are superior to biochemical conversion methods [13]. In the thermochemical process, biomass decomposes to biofuel, energy, and chemicals by heat and chemical processes [45]. Thermochemical conversion is divided into liquefaction, pyrolysis, gasification, and carbonization. Among them, hydrogen production can be obtained via pyrolysis and gasification. Hence, two of these methods are explained in detail below.

2.2.1 Pyrolysis

Pyrolysis is carried out in the absence of air or oxygen and temperature between 400 °C and 1000 °C. Based on the rate of heating, residence time and process temperature, particle size, and composition of the waste, the product of pyrolysis namely char, oil, and gaseous compounds can change [46, 47]. Depending on the heating rate and residence time, pyrolysis is termed slow pyrolysis, intermediate pyrolysis, fast pyrolysis, and flash pyrolysis. Slow pyrolysis is carried out for 25–35 h at a heating rate of 5–10 °C/min in the temperature range of 200–550 °C. The main target of slow pyrolysis condition is solid fuel (char) while less amount of liquid and gaseous product also generated. Intermediate pyrolysis performs for 10 min at a heating rate of 200–300 °C/min in the temperature range of 300–450 °C. 35–50% bio-oil, 25–40% biochar, and 20–30% gas products are obtained by the process [35, 48, 49]. In the fast pyrolysis process, vapor, tar, and char are obtained by converting the biomass. Bio-oil which is the main product is obtained through condensing of condensable vapor. The process yields 75% liquid, %12 solid, and %13 gas in a short residence time (1 s) [48]. Similar to fast pyrolysis, the highest bio-oil content up to 75% is achieved via flash pyrolysis [13]. Further steam reforming is applied to produce hydrogen. Not only recovering hydrogen from gaseous products but also, oily products can be used for hydrogen production [50].

To decrease reaction temperature, enhance the product yield and quality, decrease the secondary reaction, and enable higher energy efficiency and minimize cost, advanced pyrolysis techniques that involve catalytic pyrolysis, microwave pyrolysis, vacuum pyrolysis, solar pyrolysis, co-pyrolysis, and carbon dioxide pyrolysis are improved.

2.2.2 Gasification

Thanks to producing both hydrogen and syngas, gasification is considered the most attractive method among the other hydrogen production methods [51]. The gasification process is carried out at a higher temperature (>700 °C), in the presence of gasifying agents such as oxygen, air, steam, or a mixture of them [52, 53]. Syngas or product gas that consists of hydrogen, carbon monoxide, carbon dioxide, methane, and lighter hydrocarbons such as ethane and propane, and heavier hydrocarbons like tars at a temperature higher than 700 °C is produced through the gasification process [54]. The gasification process includes important reactions that are drying, pyrolysis, oxidation, and reduction reactions [55].

Quality and quantity of syngas have affected various parameters such as feedstock, reactor type, gasifying agent, catalyst usage, operating variables including temperature, resistance time, steam-to-biomass ratio, equivalence ratio, etc. [56–59]. Among them, gasifying agents play a crucial role in syngas composition. To give an illustration, air is the widely used gasifying agent due to being cheap but the calorific values of syngas and the overall efficiency of the gasification process are low due to high nitrogen content in syngas [60]. In order to prevent bottlenecks, steam as the gasifying agent has gained attraction. Steam gasification is considered an effective method for generating hydrogen in terms of the highest yield of hydrogen and having less environmental impact [49].

Supercritical water gasification (SCWG) is one of the thermochemical hydrogen production methods that use supercritical water as a gasifying agent. 374.1 °C and 22.1 MPa is a critical point of water that shows excellent properties including low viscosity, low dielectric constant, improved thermal conductivity, superior dissolving, and lower mass transfer resistance [61]. Because of the above-mentioned properties, SCWG has gained more attention recently [62].

The process is carried out at higher pressure while lower temperatures (600 °C) than the dry gasification process. One of the advantages of SCWG is that it is possible to apply to wet biomass without any drying pre-treatment [63]. In these conditions, biomass is converted into hydrogen-rich syngas through a series of reactions involving steam reforming reaction, water–gas conversion, and methanation [64]. Thanks to the unique properties of supercritical water, the high part of biomass is solved and a homogeneous mixture occurs [65].

3 Case Study for Hydrogen Production via Gasification Process: Comparison of EU and Turkey Projects

In the case study, the projects undertaken in EU countries and Turkey were examined using databases created by the country. Projects in EU countries were searched using the cordis.europa.eu database specifically for projects covering Horizon 2020 and FP7 for both "gasification" and ("gasification" AND "hydrogen") terms. The projects in Turkey were examined for "gasification" using trdizin.gov.tr data base.

3.1 Results and Discussions

The reviewed projects conducted in the EU and Turkey are given as subheadings.

3.1.1 Results of the Projects Carried Out by the European Union Member Countries

The Community Research and Development Information Service (CORDIS) is the European Commission's primary source of results that are funded by the EU's framework programs for research and innovation (FP1 to Horizon 2020) [66]. In this study, a project search was carried out using the cordis.europa.eu database for the terms "gasification" and "hydrogen" within Horizon 20 and FP7 projects. In Table 1, projects that include both terms "gasification" and "hydrogen" are shown.

In the above-listed projects, it is seen that the aim is to produce multiple products such as hydrogen, chemicals, syngas, alcohol, and methanol. Moreover, reactor and system development, upgrading of gas cleaning, and generating efficient catalysts were priorities of the projects. On the other hand, the feasibility and commercializing potential of developed technologies were also considered within the EU framework. In addition, projects aimed at strengthening the technological and scientific capacity in order to increase knowledge about hydrogen have been supported.

Within the scope of the Horizon 20 and FP7 projects, the projects for "gasification" in the last decade have been examined using the cordis.europa.eu database. 84 gasification projects were found according to the search results. Among them, 34 projects were found as more suitable for the scope of the study that including gasification technology. In Table 2, projects related to gasification are shown.

In the above-listed gasification projects, it can be seen that production of different types of fuel, chemicals, heat and power generation, increasing overall efficiency, reduction of emission or achieving almost zero-emission, and catalyst development for tar elimination were targeted. Especially, it can be seen that projects that included CO_2 capture and CHP are widely investigated as means of decarbonizing energy and fuel requirements.

3.1.2 Results of the Projects Carried Out by Turkey

The Scientific and Technological Research Council of Turkey is one of the relevant organizations of the Ministry of Industry and Technology with the objection to promoting science and technology. Table 3 shows the project data obtained by scanning the word "gasification" from the TÜBİTAK database of "trdizin.gov.tr".

It can be concluded from Table 3 that several aspects of the topic have been studied including the development of a new gasification system and catalyst effect on the gasification process. Most of the projects were performed at a lab scale and concluded to have the potential to scale up to industry. However, scaling up to industry is still insufficient. Although gasification technology is used in Turkey by some energy firms, it is just applied for heating and generating electricity. For this purpose, the biggest players in the industry that are namely ITC Türkiye, Biotrend, Oltan ve Kökeoğlu, ISTAÇ, Ortadoğu Enerji, Bientaş, Gülsan Holding, İstanbul Enerji, Bemka Enerji, Panab Enerji, Global Yatırım Holding, Mutlular Enerji, Eren Holding, Era Çevre Teknolojileri, Arel Enerji carry out their activities [67].

Table 1 The project that includes the term "gasification" and "hydrogen"

Project name/id	Date/country	Scale	Aim
Waste2H2/ 952,593	2021–2023/ Portugal	–	Improving the scientific and technological capacity of IPPortalegra in the cleaning and purification of thermal gasification syngas to generate hydrogen
SYNERGY/ 744,197	2016–2017/ Italy	Small plant	Converting industrial by-products into fuel gas (syngas) through gasification
UNIfHY/ 299,732	2012–2016/ Italy	Pilot & industrial scale	Enhancing materials and reactors for the integration of advanced biomass steam gasification and syngas purification processes
FLEXI-PYROCAT/ 643,322	2015–2018/ UK		Developing a technology to treat waste plastics to produce high-value-added products including hydrogen, carbon nanotubes, chemicals, or gasoline
MefCO2/ 637,016	2014–2019/ Spain	Intermediate scale	Methanol production via H_2 is produced from renewable energy and emitted CO_2
Ambition/ 731,263	2016–2016/ Norway		Enhancing the conversion process regarding both material and energy efficiency, and decreasing capital and operation costs to enable use in the European Energy system
WASTE2GRIDS/ 826,161	2019–2020/ Switzerland		Investigating the long duration techno-economic feasibility of W2G plants
H2-IGCC/ 239,249	2009–2014/ Belgium		Within the integrated approach, allowing deployment of high-efficiency gas turbines in competitive IGCC plants with CCS technology
ROMEO/680395	2015–2019/ Germany	Lab-scale	Application of ROMEO technology to the water–gas-shift process produces hydrogen from biogenic waste materials

Based on The World Bank classification, Turkey takes place in developing countries. Because of the rapid growth of industry, more energy is required. To achieve the CO_2 reduction target, Turkey should consider investing in biomass conversion technologies and step forward to the hydrogen economy. Hydrogen is not a new term for Turkey that the first initiative took place in 1992 with International Centre for Hydrogen Energy Technologies (UNIDO-ICHET) [68].

Table 2 EU Project in regarding gasification

Project name/id	Date/country	Scale	Aim
BLAZE/ 815,284	2019–2022	small to medium s	Advancing of small-to-medium biomass CHP that has a low price, and emitting no particulate matter
PYROGAS/ 286,244	2012–2016/ UK	Laboratory to demonstrate the scale	The objective of the PYROGAS project is, through a program of laboratory and demonstration-scale testing, to integrate a pyrolysis process with a gasification process to form the complete PYROGAS concept from feedstock(s) to the delivery of power and heat, and to demonstrate its successful operation on a range of feedstocks
GASPRO-BIO-WASTE/ 261,911	2011–2012/ Netherlands	Lab-scale	Developing a new analyzer involving an in situ high-temperature sensor that enables it to work within a wide range of temperatures and pressure for lab-scale gasification
ICOMFLUID/ 312,261	2012–2016/ UK	–	– Training and exchange of a combination of Early Stage Career and Established Researchers for the latest techniques including the full kind of thermal conversion to produce clean energy
FASTCARD/ 604,277	2014–2017/ Norway	–	– Enhancing an original “rational design” of nano-catalysts to control much better; optimized based upon improved characterization methods – Developing industrially relevant, insightful Downscaling methodologies
TWIN-PEAKS/ 951,308	2020–2023/ Lithuania	–	Establishing a joint research strategy regarding Waste-to-energy (WtE)
TOPS/ 608,517	2013–2017/ UK	–	– Decreasing the requirement of on-site CO_2 storage capacity – Maximizing the economic yield via producing synthetic gas – Power generation
ENV-BIO/ 318,927	2013–2016/ Spain	–	Development of enhanced thermochemical processes for the energy evaluation of biomass and assessment

(continued)

Table 2 (continued)

Project name/id	Date/country	Scale	Aim
RENEGAS/ 699,122	2015–2015/ UK	–	Developing a feasibility study regarding Refgas system to commercialize it
PULP AND FUEL/ 818,011	2018–2022/ France		Benefiting both wet gasification and dry gasification to develop a simple and durable fuel synthesis process
PlasmaPower/ 735,818	2016–2016/ Spain	Small scale	– Including the technical, commercial, and financial feasibility of a technology platform for the highly efficient conversion of waste biomass feedstock into energy via hydro-catalytic plasma gasification
GAREP/ 673,311	2015–2015/ Finland	Small to medium scale	Developing a CHP unit using the GasEK gasification reactor that enables using local biomass
GICO/ 101,006,656	2020–2024/ Italy	Small to medium	Develops new materials (CO_2 capture sorbents; high-temperature inorganic removal sorbents; catalytic filter candles; membranes for oxygen separation and methanol production) and technologies
EFFIGAS/ 775,720	2017–2017/ Italy		New automatic and efficient biomass technology that converts organic waste into high-quality syngas with very low tar content
FLEDGED/ 727,600	2016–2020/ Italy		By combining the flexible sorption-enhanced gasification (SEG) process and novel sorption-enhanced DME synthesis (SEDMES) process, developing process that utilizes biomass for Bio-based dimethyl Ether (DME) production
FLEXCHX/ 763,910	2018–2021/ Finland	Large scale	Developing a multiple production concept for heat, power, and transport fuels utilizing biomass/ waste and excess renewable electricity

(continued)

Table 2 (continued)

Project name/id	Date/country	Scale	Aim
EUWaste/ 719,028	2016–2016/ Ukraine		Gasification technology that utilizes a different type of waste with minimum environmental footprint and emission, allows producing syngas to can be used in heat production
HiEff-BioPower/ 727,330	2016–2021/ Austria	Medium-scale	Development of highly yielding biomass CHP technology
FlexiFuel-SOFC/ 641,229	2015–2019/ Austria	Micro-scale	Development of new highly effective flexible micro-scale biomass CHP technology that has high electric and overall efficiency, nearly zero gaseous and PM emissions
INORGASS/ 251,663 DEMOCLOCK/ 268,112	2011–2013/ France 2011–2015/ Norway	Bench-scale Medium	Converting inorganic content of sewage sludge to end-off products through gasification for disposal purposes Indicate feasibility including the technical, economic, and environmental for Chemical Looping Combustion in a large-scale power plant – Converting gasified hydrocarbons to energy
CERESIS/ 101,006,717	2020–2024/ Greece		– Via phytoremediation land decontamination is supplied and producing clean biofuels from growing energy crops – Supercritical Water Gasification & Fast Pyrolysis are optimized for producing clean biofuel
COMSYN/ 727,476	2017–2021/ Finland	Pilot-scale	Converting diverse raw materials to transport fuel through gasification and FT technology
COSYNAT/ 674,065	2015–2015/ Spain	Small scale	Take COSYNAT technology from TRL 6 to TRL 9 and commercialize it
CLARA/ 817,841	2018–2022/ Germany	From lab scale to pilot scale	– Producing liquid fuel from biomass via combining chemical looping gasification (CLG) and FT technology
BECOOL/ 744,821	2017–2022/ Italy		Providing highly efficient, sustainable, and affordable value chains, to develop technologies and tools for the production of advanced biofuels in Europe

(continued)

Table 2 (continued)

Project name/id	Date/country	Scale	Aim
Heat-To-Fuel/ 764,675	2017–2021/ Austria		Producing 2nd generation biofuel through combining gasification, FT(Fischer–tropsch), and aqueous phase reforming technologies
BIOBOOST/ 282,873	2012–2015/ Germany		Optimizing de-central biomass conversion, and utilizing in small-scale combined heat and power (CHP) plants or in large-scale applications to synthesize transportation fuel and chemicals
BRISK II/ 731,101	2017–2022/ Sweden	Micro- and macro-scale	Establishing a center of perfection in the field of varied order of generation biofuels through the combining of leading European research centers
BioSFerA/ 884,208	2020–2021/ Greece	Lab and piot scale	Gasifying biogenic residues and wastes to produce biofuels for use in aviation and maritime transport
BIOFUEL/ 247,550	2010–2013/ Turkey		Co-pyrolysis and co-gasification of plastic with biomass to produce bio-fuel and chemicals

Source cordis.europa.eu

However, petroleum, coal, and natural gas are fossil fuels used in Turkey as the main energy source [69]. In the context of bioethanol production, 12 plants produce bioethanol in Turkey. Just only 3 plants still produce bioethanol for transport fuel even if 8 of the total plants can produce bioethanol [70]. Until 2011, only 2% of ethanol blended into gasoline [71] while after 2014 this became at least 3%. Given the price of gasoline, it is quite high in Turkey hence it is required to consider using bioethanol instead of gasoline [71]. This situation is valid for biodiesel, due to the lack and delay in the rules and regulations. It is estimated that the requirement of 0.3 million m^3 bioethanol/annum and 1,4 million m^3 biodiesel/annum would be by the year 2023, however, the biofuel target does not take place on the Vision 2023 energy agenda and the only goal concerning biomass just takes part on electricity generation via combustion [72].

To summarize, academia shows a strong effort to develop technologies that include alternative fuels however current policies do not meet this effort. Instead of carbon-intense coal or petroleum fuel, a transition to carbon–neutral alternative fuels involving, biofuel, hydrogen, and methanol is required, which should be supported by relevant policies.

Table 3 Turkey projects regarding gasification

Project name	Due date	Scale	Aim
"Mcwgasifier" Design, Production and Performance Evaluation of a Laboratory Scale Test Setup	2018	Lab	Development of microwave plasma gasification system to produce energy from different kinds of wastes, and performing of operational-process performance
Hydrogen and Biochar Production by Dry Fermentation from Organic Fractions of Urban Solid Wastes	2018	Lab-scale	Biochar production from the organic fractions of urban solid wastes and the residue of the dry fermentation system is called digested, and thermochemical H_2 production from biochar production by gasification method
Determination of Physical and Chemical Stability of Heterogeneous Catalysts in Reactions in Supercritical Water Atmosphere	2017	Lab-scale	The thermal and hydrothermal stability of various oxide catalysts, produced commercially and prepared by sol–gel and flame spray pyrolysis methods with nano and microcrystalline sizes, were experimentally investigated in the gasification of formaldehyde and glucose in SCW
Proper Thermal Method of Refuse Derived Fuels and Evaluation of the System Modeling Using Artificial Neural Networks	2016	Lab-scale	Development of modeling of a thermal system using "Artificial neural networks"
Development of Underground Coal Gasification Technology for Coals of Our Country	2016	Lab-scale	Development of underground coal in-situ gasification system
Hydrogen And Biofuel Production From Olive Land Water	2012	Lab-scale	Supercritical water gasification of olive land water to produce hydrogen and biogas

Source trdizin.gov.tr

4 Conclusion and Future Perspective

Decarbonizing the transport sector is essential to reduce CO_2 emissions and prevent the adverse impacts of global warming while maintaining energy security. To enhance the conversion of biomass/waste to biofuel, assessments on biomass/waste potential including quality, quantity, production, transportation, management, and appropriate technology are required. Furthermore, to reach the aim of the bio-based economy, large-scale production is necessary. However, except for some countries including the EU, academic studies and investment in biomass conversion technologies are left behind on a laboratory scale. In order to overcome this situation, more R&D on these technologies should be required to scale up pilot-, demo-, and finally to large-scale.

Countries in the European Union try to decarbonize their energy, industry, and transport sectors. Therefore, while continuing their scientific studies for the advancement of technology academically in the field of alternative fuels, they also strive to put technology into practice by strengthening academia-industry cooperation.

This study was conducted to determine gasification and hydrogen projects in the EU and Turkey. The results for projects focusing on hydrogen production from gasification and other alternative fuels are presented.

Gasification is a mature technology that enables the production of hydrogen-rich syngas. However, it requires enhancing its efficiency, clean syngas production, and reducing initial costs. When conducted projects are reviewed, it is found that in contrast to Turkey, in the case of projects carried out by EU countries, these developments are realized. The policies should be amended for the development of alternative fuel and the widening of its usage to reduce the energy dependence of Turkey. To secure its energy demand and reach determined CO_2 reduction targets, it is necessary to avoid technologies using carbon-intensive fuels. Turkey has a great biomass potential with the only agricultural waste amount of 62,206,754 tons/year [73]. This potential should be evaluated through integrated biorefinery that produces not only heat/electricity but also produce hydrogen, chemical, and high-added-value materials [74].

Acknowledgements This study is supported by the Ege University Scientific Research Projects Coordination Unit. Project number: FDK-2020-22276. All authors conceived and designed the study. All authors contributed to manuscript writing and revision. All authors contributed to the design of the figures. All authors read and approved the final manuscript. Editing of the manuscript was made by Melih Soner Çeliktas.

References

1. Atilhan S et al (2021) Green hydrogen as an alternative fuel for the shipping industry. Curr Opin Chem Eng. https://doi.org/10.1016/j.coche.2020.100668
2. Celiktas MS, Alptekin FM (2019) Conversion of model biomass to carbon-based material with high conductivity by using carbonization. Energy 188. https://doi.org/10.1016/j.energy.2019.116089
3. Romejko K, Nakano M (2017) Portfolio analysis of alternative fuel vehicles considering technological advancement, energy security and policy. J Clean Prod 142. https://doi.org/10.1016/j.jclepro.2016.09.029
4. Navas-Anguita Z, García-Gusano D, Iribarren D (2020) Long-term production technology mix of alternative fuels for road transport: A focus on Spain. Energy Convers Manage 226. https://doi.org/10.1016/j.enconman.2020.113498
5. Foorginezhad S et al (2021) Sensing advancement towards safety assessment of hydrogen fuel cell vehicles. J Power Sour. https://doi.org/10.1016/j.jpowsour.2021.229450
6. Uyan M et al (2020) Bioconversion of hazelnut shell using near critical water pretreatment for second generation biofuel production. Fuel 273. https://doi.org/10.1016/j.fuel.2020.117641

7. Nzihou A (2010) Toward the valorization of waste and biomass. Waste Biomass Valor 1(1). https://doi.org/10.1007/s12649-010-9014-x
8. Jąderko-Skubis K (2021) Production of alternative fuels from waste: assumptions for the design of new fuel recipes. Int J Sustain Eng. https://doi.org/10.1080/19397038.2020.1836063
9. Inyinbor AA, Oluyori AP, Adelani-Akande TA (2017) Biomass valorization: Agricultural waste in environmental protection, phytomedicine and biofuel production. In: Biomass volume estimation and valorization for energy. https://doi.org/10.5772/66102
10. Sadala S et al (2019) Resource recovery as alternative fuel and raw material from hazardous waste. Waste Manage Res. https://doi.org/10.1177/0734242X19854124
11. Aziz M, Darmawan A, Juangsa FB (2021) Hydrogen production from biomasses and wastes: A technological review. Int J Hydrogen Energy. https://doi.org/10.1016/j.ijhydene.2021.07.189
12. Madadian E, Amiri L, Lefsrud M (2020) Thermodynamic analysis of wood pellet gasification in a downdraft reactor for advanced biofuel production. Waste Biomass Valor 11(7). https://doi.org/10.1007/s12649-019-00663-4
13. Osman AI et al (2021) Conversion of biomass to biofuels and life cycle assessment: A review. Environ Chem Lett. https://doi.org/10.1007/s10311-021-01273-0
14. Stančin H et al (2020) A review on alternative fuels in future energy system. Renew Sustain Energy Rev 128. https://doi.org/10.1016/j.rser.2020.109927
15. Brynolf S et al (2018) Electrofuels for the transport sector: A review of production costs. Renew Sustain Energy Rev. https://doi.org/10.1016/j.rser.2017.05.288
16. Lester MS, Bramstoft R, Münster M (2020) Analysis on electrofuels in future energy systems: A 2050 case study. Energy 199. https://doi.org/10.1016/j.energy.2020.117408
17. Wang A, Austin D, Song H (2019) Investigations of thermochemical upgrading of biomass and its model compounds: Opportunities for methane utilization. Fuel. https://doi.org/10.1016/j.fuel.2019.03.015
18. Parra D et al (2019) A review on the role, cost and value of hydrogen energy systems for deep decarbonisation. Renew Sustain Energy Rev. https://doi.org/10.1016/j.rser.2018.11.010
19. Okolie JA, Patra BR et al (2021) Futuristic applications of hydrogen in energy, biorefining, aerospace, pharmaceuticals and metallurgy. Int J Hydrogen Energy. https://doi.org/10.1016/j.ijhydene.2021.01.014
20. Alptekin FM, Çeliktaş MS (2020) Renewable energy and its future from hydrogen energy perspective. In: 4th International congress on economics finance and energy. Niğde, Turkey.
21. Edwards PP et al (2008) Hydrogen and fuel cells: Towards a sustainable energy future. Energy Policy 36(12). https://doi.org/10.1016/j.enpol.2008.09.036
22. Hosseini SE, Wahid MA (2016) Hydrogen production from renewable and sustainable energy resources: Promising green energy carrier for clean development. Renew Sustain Energy Rev. https://doi.org/10.1016/j.rser.2015.12.112
23. Navas-Anguita Z et al (2021) Revisiting the role of steam methane reforming with CO2 capture and storage for long-term hydrogen production. Sci Total Environ 771. https://doi.org/10.1016/j.scitotenv.2021.145432
24. Velazquez Abad A, Dodds PE (2020) Green hydrogen characterisation initiatives: Definitions, standards, guarantees of origin, and challenges. Energy Policy 138. https://doi.org/10.1016/j.enpol.2020.111300
25. Navas-Anguita Z, García-Gusano D, Iribarren D (2019) A review of techno-economic data for road transportation fuels. Renew Sustain Energy Rev. https://doi.org/10.1016/j.rser.2019.05.041
26. Spyra H, Salmhofer HJ (2016) The politics of decarbonisation—A case study. Transp Res Procedia 14:4050–4059. https://doi.org/10.1016/j.trpro.2016.05.502
27. European Commission (2014) Directive 2014/94/EU—Deployment of alternative fuels infrastructure. Off J Eur Union L307(October):20. Available at https://eur-lex.europa.eu/legal-content/EN/TXT/?uri=CELEX:32014L0094
28. Kowalska-Pyzalska A, Kott J, Kott M (2020) Why Polish market of alternative fuel vehicles (AFVs) is the smallest in Europe? SWOT analysis of opportunities and threats. Renew Sustain Energy Rev. https://doi.org/10.1016/j.rser.2020.110076

29. Viesi D, Crema L, Testi M (2017) The Italian hydrogen mobility scenario implementing the European directive on alternative fuels infrastructure (DAFI 2014/94/EU). Int J Hydrogen Energy 42(44):27354–27373. https://doi.org/10.1016/j.ijhydene.2017.08.203
30. Adams P et al (2018) Chapter 8—Biomass conversion technologies BT—Greenhouse gas balances of bioenergy systems. In: Greenhouse gas balances of bioenergy systems. Elsevier, Amsterdam
31. Garba A (2021) Biomass conversion technologies for bioenergy generation: An introduction. In: Biotechnological applications of biomass. IntechOpen. https://doi.org/10.5772/intechopen.93669
32. Liu Y et al (2019) Cascade utilization of lignocellulosic biomass to high-value products. Green Chem. https://doi.org/10.1039/c9gc00473d
33. Dou B et al (2019) Hydrogen production from the thermochemical conversion of biomass: Issues and challenges. Sustain Energy Fuels. https://doi.org/10.1039/c8se00535d
34. Hosseini SE et al (2015) A review on biomass-based hydrogen production for renewable energy supply. Int J Energy Res. https://doi.org/10.1002/er.3381
35. Varma AK, Shankar R, Mondal, P (2018) A review on pyrolysis of biomass and the impacts of operating conditions on product yield, quality, and upgradation. In: Recent advancements in biofuels and bioenergy utilization. https://doi.org/10.1007/978-981-13-1307-3_10.
36. Verma M et al (2012) Biofuels production from biomass by thermochemical conversion technologies. Int J Chem Eng. https://doi.org/10.1155/2012/542426
37. Kalinci Y, Hepbasli A, Dincer I (2009) Biomass-based hydrogen production: A review and analysis. Int J Hydrogen Energy. https://doi.org/10.1016/j.ijhydene.2009.08.078
38. Lepage T et al (2021) Biomass-to-hydrogen: A review of main routes production, processes evaluation and techno-economical assessment. Biomass Bioenerg. https://doi.org/10.1016/j.biombioe.2020.105920
39. Karim A et al (2021) Technical difficulties of mixed culture driven waste biomass-based biohydrogen production: Sustainability of current pretreatment techniques and future prospective. Renew Sustain Energy Rev. https://doi.org/10.1016/j.rser.2021.111519
40. Mishra P et al (2019) Outlook of fermentative hydrogen production techniques: An overview of dark, photo and integrated dark-photo fermentative approach to biomass. Energy Strat Rev. https://doi.org/10.1016/j.esr.2019.01.001
41. Bhatia SK et al (2021) Renewable biohydrogen production from lignocellulosic biomass using fermentation and integration of systems with other energy generation technologies. Sci Total Environ 765:144429. https://doi.org/10.1016/j.scitotenv.2020.144429
42. Liu Y et al (2019) Recent developments of hydrogen production from sewage sludge by biological and thermochemical process. Int J Hydrogen Energy. https://doi.org/10.1016/j.ijhydene.2019.06.044
43. Hitam CNC, Jalil AA (2020) A review on biohydrogen production through photo-fermentation of lignocellulosic biomass. Biomass Conversion and Biorefinery. https://doi.org/10.1007/s13399-020-01140-y
44. Łukajtis R et al (2018) Hydrogen production from biomass using dark fermentation. Renew Sustain Energy Rev. https://doi.org/10.1016/j.rser.2018.04.043
45. Okolie JA, Mukherjee A et al (2021) Next-generation biofuels and platform biochemicals from lignocellulosic biomass. Int J Energy Res. https://doi.org/10.1002/er.6697
46. Kalyani KA, Pandey KK (2014) Waste to energy status in India: A short review. Renew Sustain Energy Rev. https://doi.org/10.1016/j.rser.2013.11.020
47. Lombardi L, Carnevale E, Corti A (2015) A review of technologies and performances of thermal treatment systems for energy recovery from waste. Waste Manage 37. https://doi.org/10.1016/j.wasman.2014.11.010
48. Bridgwater AV (2012) Review of fast pyrolysis of biomass and product upgrading. Biomass and Bioenergy 38. https://doi.org/10.1016/j.biombioe.2011.01.048
49. Pandey B, Prajapati YK, Sheth PN (2019) Recent progress in thermochemical techniques to produce hydrogen gas from biomass: A state of the art review. Int J Hydrogen Energy. https://doi.org/10.1016/j.ijhydene.2019.08.031

50. Parthasarathy P, Narayanan KS (2014) Hydrogen production from steam gasification of biomass: Influence of process parameters on hydrogen yield—A review. Renew Energy. https://doi.org/10.1016/j.renene.2013.12.025
51. Peng WX et al (2017) Hydrogen and syngas production by catalytic biomass gasification. Energy Convers Manage 135:270–273. https://doi.org/10.1016/j.enconman.2016.12.056
52. Ayodele BV et al (2019) A mini-review on hydrogen-rich syngas production by thermo-catalytic and bioconversion of biomass and its environmental implications. Front Energy Res. https://doi.org/10.3389/fenrg.2019.00118
53. Cerone N et al (2017) Pilot plant air-steam gasification of nut shells for syngas production. In: European Biomass Conference and Exhibition Proceedings, pp 843–846
54. Ong HC et al (2019) Catalytic thermochemical conversion of biomass for biofuel production: A comprehensive review. Renew Sustain Energy Rev. https://doi.org/10.1016/j.rser.2019.109266
55. Udomsirichakorn J, Salam PA (2014) Review of hydrogen-enriched gas production from steam gasification of biomass: The prospect of CaO-based chemical looping gasification. Renew Sustain Energy Rev. https://doi.org/10.1016/j.rser.2013.10.013
56. Chen G et al (2020) Hydrogen-rich syngas production from municipal solid waste gasification through the application of central composite design: An optimization study. Int J Hydrogen Energy. https://doi.org/10.1016/j.ijhydene.2020.09.118
57. Li W et al (2014) Investigation of hydrogen production using wood pellets gasification with steam at high temperature over 800 °C to 1435 °C. In: International Journal of Hydrogen Energy. Elsevier Ltd, pp 5580–5588. https://doi.org/10.1016/j.ijhydene.2014.01.102
58. Nzihou A, Stanmore B, Sharrock P (2013) A review of catalysts for the gasification of biomass char, with some reference to coal. Energy 58. https://doi.org/10.1016/j.energy.2013.05.057
59. Tian T et al (2017) Effects of biochemical composition on hydrogen production by biomass gasification. Int J Hydrogen Energy 42(31):19723–19732. https://doi.org/10.1016/j.ijhydene.2017.06.174
60. Xu CC et al (2018) Biomass energy. Compr Energy Syst 1–5:770–794. https://doi.org/10.1016/B978-0-12-809597-3.00121-8
61. Kıpçak, E., Akgün, M. (2015). Hydrogen Production by Supercritical Water Gasification of Biomass. In: Fang, Z., Smith, Jr., R., Qi, X. (eds) Production of Hydrogen from Renewable Resources. Biofuels and Biorefineries, vol 5. Springer, Dordrecht. https://doi.org/10.1007/978-94-017-7330-0_7
62. Zhang Y et al (2019) Hydrogen production through biomass gasification in supercritical water: A review from exergy aspect. Int J Hydrogen Energy. https://doi.org/10.1016/j.ijhydene.2019.01.151
63. Rodriguez Correa C, Kruse A (2018) Supercritical water gasification of biomass for hydrogen production—Review. J Supercrit Fluids 133:573–590. https://doi.org/10.1016/J.SUPFLU.2017.09.019
64. Cao L et al (2020) Biorenewable hydrogen production through biomass gasification: A review and future prospects. Env Res 186:109547. https://doi.org/10.1016/J.ENVRES.2020.109547
65. Safari F et al (2015) Hydrogen and syngas production from gasification of lignocellulosic biomass in supercritical water media. Int J Recycl Org Waste Agric 4(2). https://doi.org/10.1007/s40093-015-0091-5
66. Must Ü (2020) Priorities for social and humanities projects based on text analysis. J Data Inf Sci 5(4). https://doi.org/10.2478/jdis-2020-0036
67. Çolakogu M Aslan, Kumdereli, SE (2021) Biyokütle ve Biyoenerji Sektörüne Genel Bakış. Available at https://www.pwc.com.tr/tr/gundem/beceri-kazandirma.html
68. Celiktas MS, Kocar G (2010) Hydrogen is not an utopia for Turkey. Int J Hydrogen Energy 35(1). https://doi.org/10.1016/j.ijhydene.2009.10.067
69. Bayrakci Ozdingis AG, Kocar G (2018) Current and future aspects of bioethanol production and utilization in Turkey. Renew Sustain Energy Rev. https://doi.org/10.1016/j.rser.2017.06.031
70. Türközü B, Özen DN, Altinişik K (2010) The energy efficiency utilization of the cumra sugar factory with exergy analysis. Isı Bilimi ve Tekniği Dergisi 30(2)

71. Bayrakci AG, Koçar G (2014) Second-generation bioethanol production from water hyacinth and duckweed in Izmir: A case study. Renew Sustain Energy Rev. https://doi.org/10.1016/j.rser.2013.10.011
72. Melikoglu M (2017) Vision 2023: Status quo and future of biomass and coal for sustainable energy generation in Turkey. Renew Sustain Energy Rev. https://doi.org/10.1016/j.rser.2017.03.005
73. T.R. Energy and Natural Resources Ministry (2021) Turkey biomass energy potential Atlas. Available at https://bepa.enerji.gov.tr/
74. Uyan M, Alptekin FM, Çeliktaş MS (2017) Biorefinery concept: Current status and future prospects. Konya Turkey. Available at https://icente.selcuk.edu.tr/uploads/files2/ICENTE_proceeding_V3.pdf

Targeted Synthesis of Hydrocarbon Fuels and Fuel Oxygenates by Catalytic Conversion of Biomass Components

Navya Subray Bhat, Saikat Dutta, and Girdhar Joshi

Abstract Biomass has been projected as a commercially tenable, renewable alternative to fossilized sources such as petroleum for producing liquid transportation fuels. This chapter provides an overview of the catalytic synthesis of drop-in hydrocarbon fuels and oxygenated fuel additives starting from various biomass components. The catalytic conversion of biomass affords high selectivity and precise molecular control over the fuel candidates under reagent-economic and energy-efficient conditions. Emphasis has been given to the synthetic strategies and process details for the designed synthesis of aviation, gasoline, and diesel-range fuels. Details of the feedstock, reaction conditions, and catalysts employed have also been incorporated. The major biomass components have been discussed separately, and their catalytic conversion into specific fuels has been elaborated. This chapter will enable the readers to acquire inclusive knowledge about this field of research, appreciate the accomplishments, identify the bottlenecks, and propose the way forward.

1 Introduction

The standard of living in a country may be correlated with the per capita consumption of fossilized carbon (e.g., petroleum) [1]. A large fraction of a drum of crude oil is utilized to produce hydrocarbon-based liquid transportation fuels [2, 3]. However, anthropogenic resources have limited reserves, and they are concentrated in specific geographic locations. Our overdependence on petroleum for transportation fuels and chemicals has caused serious geopolitical, economic, and environmental consequences [4]. In many futuristic scenarios, biomass (fresh and waste) fits seamlessly in the renewable energy landscape as a sustainable alternative to petroleum for making

N. S. Bhat · S. Dutta (✉)
Department of Chemistry, National Institute of Technology Karnataka (NITK), Surathkal, Mangalore, Karnataka 575025, India
e-mail: sdutta@nitk.edu.in

G. Joshi
Department of Chemistry, Sridev Suman Uttarakhand University Campus (Govt. P. G. College), Gopeshwar, Chamoli, Uttarakhand 246401, India

A. Khan et al. (eds.), *Waste to Biofuel Technology*, Sustainable Materials and Technology, https://doi.org/10.1007/978-981-97-4561-6_3

transportation fuels [5]. Biomass is geographically diverse, available in bulk quantities, inexpensive (especially the waste biomass), sustainable, and considered less polluting [6]. Nature took millions of years to transform biomass into hydrocarbon-rich petroleum by chemical reactions under elevated temperature and pressure. The primary technological challenge for scientists to produce biofuels is to accomplish the same in a controlled environment within minutes to hours. Biomass comprises biopolymers such as cellulose, hemicellulose, and lignin. In addition, some biomasses contain notable quantities of extractives (e.g., triglyceride, protein) and inorganic compounds [7]. The composition of biomass varies with the nature of biomass, climatic conditions, and even storage parameters. Cellulose, the single most abundant biopolymer available in nature, is made of glucose units linearly stitched by the β-1,4-glycosidic bond. Hemicellulose is a non-linear polymer made of hexose (e.g., glucose) and pentose (e.g., xylose) sugars. Lignin is a complex biopolymer consisting of substituted phenolic units connected by methylene, ether, and carbonyl groups. The heavily oxygenated biopolymers and extractives must be deoxygenated partially or fully, and their structure simplified for biofuels synthesis [8]. The choice of biomass for biofuels synthesis is paramount. The first-generation biofuels used starchy biomass (for bioethanol) and edible oils (for biodiesel). However, it became evident that diverting food resources for producing biofuels could worsen food scarcity in vulnerable parts of the world. Besides, the production of biofuels from edible biomass necessitates significant materials (e.g., fertilizers, pesticides) and energy (e.g., irrigation, harvesting) input, many of which utilize petroleum in the first place [9]. Therefore, first-generation biofuels cannot be labeled carbon-neutral in the true sense. The second and third-generation biofuels use lignocellulosic biomass and non-edible oil-bearing biomass. These feedstocks have high productivity, do not compete with the human food chain, require minimal energy and materials during production, and are often considered waste [10]. Freshwater and marine algae have also shown promise for synthesizing third-generation biofuels (Fig. 1). There are three major biomass valorization pathways: thermochemical, chemocatalytic, and biotechnological [11]. The thermochemical strategies primarily use heat to depolymerize and deoxygenate the biopolymers into smaller organics with limited oxygen content. The pyrolytic liquefaction of biomass in the presence or absence of a catalyst leads to a mixture of liquid compounds (termed as bio-oil), which may be used as fuel after suitable upgrading [12]. The gasification technique combusts biomass in a limited oxygen supply to produce a gaseous mixture primarily consisting of CO and H_2 (syngas). The molar ratio between CO and H_2 is adjusted and converted into liquid hydrocarbon fuels using the Fischer-Tropsch process [13]. The biotechnological processes use microorganisms (native or genetically engineered) to deconstruct the biopolymers into small organics (e.g., ethanol, 1-butanol), which can be used as fuel directly or after appropriate synthetic modifications [14]. The chemocatalytic value addition of biomass has received particular interest since the processes are fast, selective, energy-efficient, do not require live organisms, and could potentially be integrated into the existing petrorefinery infrastructure [15]. Catalysis conforms to many of the principles of green chemistry and is a crucial strategy for biofuels production. A wealth of data have been generated on various catalysts for biofuels

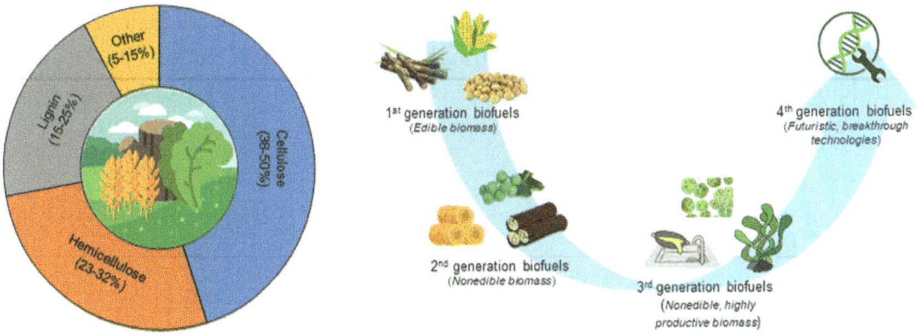

Fig. 1 Composition of lignocellulosic biomass (*left*) and the evolution of biofuels (*right*)

synthesis [16]. Heterogeneous catalysts are preferred over their homogeneous counterpart due to the axiomatic advantage of the former toward separation and recovery [17]. In recent years, nanocatalysts and single-atom catalysts have been increasingly used owing to their high catalytic activity under relatively mild conditions, bridging the gap between homogeneous and heterogeneous catalysts [18, 19]. Biofuels can be classified into two main categories, namely, drop-in fuels and fuel additives. The drop-in biofuels are essentially hydrocarbons, which are structurally identical to those originating from petroleum and can be used as standalone fuels [20]. Fuel additives are typically oxygenated hydrocarbons containing at least one oxygen atom in their moiety. The fuel additives are often blended with hydrocarbon fuels (e.g., gasoline, diesel) for complete combustion and lesser emission [21]. The targeted synthesis of fuels allows precise control over their molecular structures, makes their downstream processing largely redundant, and the process economically competitive. Besides, the target-oriented synthesis of biofuels ensures reproducible molecular composition of the fuel mixture, predictable physicochemical properties, and combustion characteristics [22]. This chapter describes the targeted synthesis of hydrocarbon fuels and fuel additives starting from various biomass components using chemocatalytic pathways. The designed synthesis of hydrocarbon drop-in and oxygenated fuel additives from various biomass components following catalytic routes is emphasized. The details of organic transformations, catalysts employed, and reaction conditions applied are also elaborated.

2 Fuels from Triglyceride

Triglyceride in vegetable oils and animal fats are crucial extractives in biomass, having several established and emerging markets [23]. Although oil-bearing biomass represents only a minor fraction of all available biomass, triglyceride is relatively easy to isolate and valorize for being non-polymeric. Oils and fats have been used as domestic fuels (for light and heat) even before the commercial adoption of petroleum due to their high calorific value and other favorable combustion characteristics.

Triglycerides are triesters of glycerol with fatty acids. The fatty acids are saturated, monounsaturated, or polyunsaturated straight-chain carboxylic acids with different chain lengths [24]. The fuel applications of triglycerides will be discussed in the following sections, especially those coming from the fatty acid component.

2.1 *Biodiesel*

Biodiesel has gained significant attention as one of the most promising sources derived from renewable sources with high quality, which allows the replacement of petrodiesel without engine modifications [25–29]. As an alternative to diesel fuel, biodiesel is well recognized for its renewability, sulfur-free and oxygenated nature, sustainability, biodegradability, non-toxic, high flash-point, and eco-friendly nature compared to conventional diesel fuel [27, 28]. Biodiesel can be easily blended with conventional diesel because the blended diesel demonstrates similar fuel characteristics to conventional diesel but with a significant reduction in harmful emissions [30, 31]. Vegetable oils (edible and non-edible), algal oils, animal fats, waste cooking oils, and microbial oils are the potential sources of feedstock used for biodiesel production. These oils on treatment with alcohol (methanol/ethanol) in the presence of acidic or basic catalysts can be transformed into biodiesel and a high-value co-product glycerol (Scheme 1) [32, 33]. Biodiesel may be prepared by using both homogeneous or heterogeneous acidic and basic catalysts. Transesterification is a reversible chemical process involving a reaction between oil and alcohol in the presence of the catalyst, a liquid acid, or a liquid base.

2.1.1 Feedstocks for Biodiesel Production

As mentioned, several feedstocks are available for biodiesel production, such as edible and non-edible oils, algal oils, animal fats, waste cooking oils, and microbial oils (Table 1).

Therefore, one has to be very ingenious while selecting a particular feedstock for biodiesel production since it affects the cost of production, composition, purity, and yields of biodiesel [29, 37, 53]. The proper selection of feedstock can be made

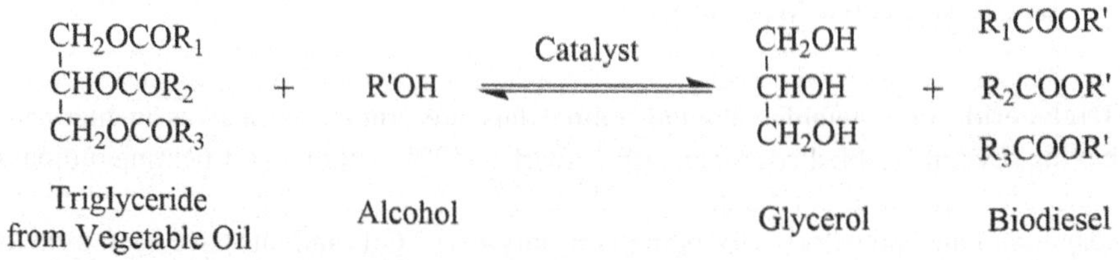

Scheme 1 Transesterification of vegetable oils

Table 1 Potential feedstocks used for first, second, third, and fourth-generation biodiesel production [26, 27, 29, 34–54]

First-generation	Second-generation	Third-generation	Fourth-generation
Soybean	Jatropha (*Jatropha curcas*)	Micro algae	Genetically modified algae
Sunflower	Karanja (*Pongamia pinnata*)	Beef tallow	Yeast
Rapeseed	Neem (*Azadirachta indica*)	Poultry fat	Fungi
Palm	Sal (*Shorea robusta*)	Pork lard	Cyanobacteria
Palm seed	Castor (*Ricinus communis*)	Waste cooking oil	Photobiological processes
Olive	Rubber Seed (*Hevea brasiliensis*)	Leather tanning waste	
Mustard	Mahua (*Madhuka longifolia*)		
Cotton seed	Jojoba (*Simmondsia chinensis*)		
Coconut	Taal (*Borassus flabellifer*)		
Rice bran	Nagchampa (*Plumeria rubra*)		
Walnut	Chinese tallow seed (*Triadeca sebifera*)		
Corn	Milk thistle (*Silybum marianum*)		
Canola	Tobacco (*Nicotiana tabacum*)		
Linseed	Soapnut (*Sapindus mukorossi*)		
Peanut	Candlenut (*Aleurites moluccanus*)		
Radish	Polanga (*Calophylluminophyllum L.*)		
Sorghum	Bottle tree (*Brachychiton rupestris*)		
	Yellow oleander (*Cascabela thevetia*)		
	Kusum (*Carthamus tinctorius*)		
	Eucalyptus oil (*Eucalyptus globulus*)		
	Linseed (*Linum usitatissimum*)		

(continued)

Table 1 (continued)

First-generation	Second-generation	Third-generation	Fourth-generation
	Mexican prickly poppy (*Argemone mexicana*)		
	Coffee ground (*Coffea arabica*)		
	Moringa (*Moringa oleifera*)		
	Cumaru (*Dipteryx odorata*)		
	Sea mango (*Cerbera odollam*)		
	Tung (*Vernicia fordii*)		
	Soapnut (*Sapindus mukorossi*)		

by analyzing various feedstock parameters, such as oil content, suitability, chemical composition, and physical properties. Various biomass sources/feedstocks reported in the literature for biofuels production are summarized in Fig. 2.

Feedstock selection for biodiesel production is also contingent on the region, availability, and the type of sources (e.g., edible, non-edible, and waste). For example, in countries like the USA and Brazil, edible feedstocks (soya oil, sunflower oil, rapeseed oil, starch, and maize) biodiesel is being used as vehicle engine fuel, and

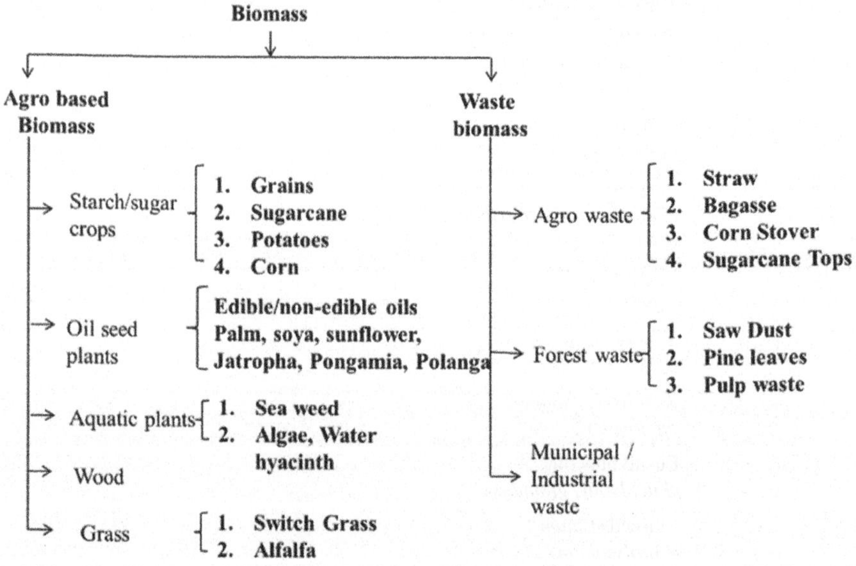

Fig. 2 Sources of biomass feedstock for biofuels production [27, 53]

this trend may continue in the coming years mainly because of sufficient availability of edible feedstocks. In Europe and tropical countries, the major sources for the production of biodiesel are palm oil and rapeseed oil [26, 27, 29, 30]. In developing countries like India and China, biodiesel production from edible sources has not received much attention since edible oil production is less than its actual demand in these countries. The use of edible oils as a source of biodiesel feedstock competes with the food supply and results in higher production costs. Therefore, the non-edible oil sources (e.g., *Jatropha*, *Karanja*, *Mahua*, *Algae*, and *Sal*) would be more appropriate for biodiesel production as it would allow compliance with the regional and ecological requirements for biofuels [27, 36, 39, 55]. Other important feedstocks such as waste cooking oil, animal fats, and microbial oils have also been used for biodiesel production [56–59]. Thus, depending on the feedstock development at various stages, biodiesel is usually classified as the first, second, third, and fourth-generation biodiesel.

Biodiesel produced from the edible feedstocks is known as first-generation biodiesel, and the respective feedstocks are first-generation feedstocks. These edible feedstocks include soyabean, rapeseed, palm, sunflower, peanut, corn, rice bran, coconut, olive, and castor oil [26, 60]. The use of such edible feedstock for biodiesel production was highly popular at the beginning of biodiesel development. However, serious concerns arose with these feedstocks, such as adequate availability of land for cultivation, biodiversity, and food supply restricted the use of such feedstocks for biodiesel production, particularly in developing countries [52, 61, 62]. In addition, the higher cost of feedstock and their growth restrictions are some other disadvantages associated with it [52, 63]. Furthermore, the extensive use of these feedstocks led to serious environmental issues, such as soil erosion, water pollution, and increased GHG emissions, especially when these crops were planted outside the traditional agriculture patterns. These constraints have forced scientists and policymakers to shift to non-edible feedstocks for biodiesel production [36].

Biodiesel production from non-edible sources is categorized as second-generation biodiesel. The second-generation feedstocks include Jatropha oil, Karanja oil, polanga oil, Neem oil, sal oil, Mahua oil, Nagchampa oil, Rubber seed oil, *Aleurites Moluccana*, salmon oil, *Madhuca longifolia*, tobacco seed, and sea mango [27, 64–66]. In comparison to the first-generation feedstocks, biodiesel production from the second-generation feedstocks offered several benefits like less production cost, more eco-friendly in nature, eradicating food inequality, and less requirement of agriculture land for cultivation. However, some disadvantages are also associated with these feedstocks, as the farmers have started cultivation of high demanding feedstocks like Jatropha, Karanja, and Jojoba in agricultural land because of the lower plant yields of these feedstocks in non-agricultural land. This has a direct influence on food production and the economy of society. The second-generation biofuel conversion technologies are also energy-intensive. Also, the additional requirement of alcohol for biodiesel production from second-generation feedstocks is another drawback associated with it. Therefore, to maintain the balance between socio-economics and biodiesel demand, researchers are paying attention to new alternate solutions, which are more economical, and easily accessible to a larger extent.

The shortcomings associated with first- and second-generation feedstocks for biodiesel production were addressed with the development of the third-generation feedstocks. The biodiesel produced from microalgae is termed the third-generation biodiesel, and the respective feedstocks are termed the third-generation feedstocks [29, 34, 52, 66]. The lesser greenhouse effect, excellent growth rate and productivity, lesser requirement of farming land, high lipid content, and lesser influence on food supply are some of the significant merits of the biodiesel produced from the microalgae. Because of their autotrophic nature, algae have a strong mitigation potential against CO_2 emissions [67]. Furthermore, due to two specific characteristics, the algae also provide various valuable substrates for producing diverse fuels [67, 68]. The first characteristic is that the algal oil can be easily converted into diesel fuel or even gasoline components [69]. Special fermentative bacteria can degrade the biomass for biobutanol production, which can be genetically manipulated for producing biobutanol, biomethane, bioethanol, vegetable oil, and jet fuel [38, 70]. The biodiesel production from algae is still in the developmental stage. The technological knowledge for large-scale production of biodiesel from algae is still limited. Therefore, the large-scale, economic, and sustainable production of algal biomass is still a major constraint for the commercial production of algal biodiesel [29]. Besides, animal fats, pork lard, beef tallow, and waste cooking oils also come under the potential feedstocks for third-generation biodiesel production [29, 35, 52]. Animal fats are the most widely used feedstock for biodiesel production among these waste oils and fats. They are preferred over first- and second-generation feedstocks because the biodiesel produced from animal fats has a high-octane number non-corrosive and has greater sustainability [34, 71]. However, the major drawback associated with these waste oil-based feedstocks is the lack of sustainable technology for the commercial production of biodiesel. Furthermore, most animal fats possess a high concentration of saturated fatty acid, which increases the transesterification complexity [35, 66].

Most recently, biodiesel can also be produced from genetically modified algae, yeast, fungi, cyanobacteria, and related photobiological processes. It is considered as the fourth-generation biodiesel. The genetically modified algae are used to enhance biodiesel production. It is achieved by using some common strategies such as improving the photosynthetic efficiency, increasing light penetration, and reducing photo-inhibition [40, 49]. The ability of these genetically modified microorganisms to convert CO_2 to fuel through a significantly improved photosynthetic process allows better mitigation of CO_2 and other greenhouse gases (GHGs) compared to the third-generation feedstocks. The biodiesel produced through the fourth-generation feedstocks also strives to have the lowest environmental impact compared to other generations. However, the study is still in the early stages of development [49]. The large-scale commercialized production of biodiesel from the fourth-generation feedstock is still in the early development stage due to insufficient biomass production and high production costs [40, 51]. Although the introduction of genetically modified algae can be seen as a promising feedstock resource to obtain the customized characteristics of desired algae; however, the major consequence will be the uncertainty associated with the species and the unknown threats that they pose to the ecosystem [71].

Although numerous feedstocks have been used for biodiesel production (Table 1), their availability in any country depends on the geographical location, climatic conditions, type of soil, and agricultural practices prevalent there [72]. The life-cycle analysis and techno-economic analysis of individual feedstock must be performed to assess the economic value of feedstock, its environmental impacts, energy economics, land availability, employment creation, logistics, and production cost [73, 74]. The production cost of biodiesel is also dependent on the end-use of the byproduct i.e. glycerol [75, 76]

2.1.2 Biodiesel Conversion Methods

Triglycerides cannot be used directly in diesel engines because of their high free fatty acid content, high viscosity, low volatility, auto-oxidation, and gum formation during storage and combustion [31, 53, 64]. Therefore, triglycerides from oil-bearing plants and animal fats are subjected to appropriate treatment for converting them into engine-compatible fuel, *i.e.,* biodiesel. Biodiesel can be prepared using thermal cracking, microemulsion, and transesterification [32, 77]. Transesterification is the most widely explored method where the triglycerides in oil or fat are reacted with an alcohol (usually methanol or ethanol used in excess) with or without a catalyst to form biodiesel (e.g., fatty acid methyl esters, fatty acid ethyl ester) and glycerol (Scheme 1). The catalysts used may be alkaline, acidic, or an enzyme. Both homogeneous and heterogeneous catalysts have been applied for the transesterification reactions. Table 2 comprises representative methods, reaction conditions, and catalysts for efficient transesterification reactions.

2.1.3 Future Perspectives of Biodiesel Technology

The biodiesel production technologies based on first- and second-generation feedstocks are well-established and commercially used for biodiesel production globally. However, some serious challenges are associated with these technologies, including the food versus fuel controversy, land availability issue, and loss of biodiversity. Biodiesel production from Jatropha seed oil (non-edible feedstock) has somewhat addressed some of these concerns. However, due to insufficient research and limited data available on the sustainable cultivation of Jatropha, sluggish participation of public and private sectors, and volatile market, this technology could not reach the commercialization stage. Other second-generation non-edible feedstocks, such as Pongamia, Polanga, Neem, and Mahua, have gained more attention for the sustainable production of biodiesel. Freshwater and marine algae have also shown tremendous potential for biodiesel production better than many first- and second-generation terrestrial feedstocks. Algae can also be used for the production of other renewable fuels, such as methane production via anaerobic digestion and bio-hydrogen via photo-biological route, with minimal environmental footprint. However, this

Table 2 Representative feedstocks, catalysts, reaction conditions, and biodiesel yield for biodiesel production

Feedstock	Catalyst (wt.%)	Alcohol to oil molar ratio	Temp. (°C)	Time (h)	Yield (%)	Refs.
Sunflower	CaO (2.0)	12:1	60	2	99.6	[78]
	CaO-based/Au nanoparticle (3.0)	9:1	n.a	3	94–97	[79]
	Mg/La (5.0)	53:1	65	0.5	100	[80]
	Mg/La (5.0)	53:1	R.T	2.2	100	[80]
Soybean	CaO (8.0)	12:1	65	3	95	[81]
	CaO-K_2O (15.0)	4.6:1	70	4	99	[82]
	$Ca(OC_2H_5)_2$ (3.0)	12:1	65	1.5	98.1	[83]
	Eu_2O_3/Al_2O_3 (10.0)	6:1	70	8	63	[84]
	KI/Al_2O_3 (2.5)	15:1	65	8	96	[85]
	KNO_3/Al_2O_3 (6.5)	15:1	65	7	87	[86]
	S-ZrO_2 sulfated zirconia (5.0)	20:1	120	1	98.6	[87]
	KOH/NaX zeolite (3.0)	10:1	65	8	85.6	[88]
	Cs-Na_2ZrO_3 (1.0)	30:1	65	0.25	98.8	[89]
	Mg–Al hydrotalcite (7.5)	15:1	65	9	67	[90]
	[a]$Ti(Pr)_4Al(Pr)_3$ (3.0)	6:1	60	2	64	[91]
	[a]NaOMe or KOMe (0.15)	6:1	45	1	100	[92]
	[a]CaO (7.0)	10:1	62	1	90	[93]
	[b]CaO (3.0)	7:1	65	1	96.6	[94]
Palm	CaO (9.0)	12:1	60	2	94–97	[95]
	CaO functionalized with Sr (5.0)	9:1	65	0.5	98.4	[96]
	[a]CaO (8.0)	9:1	60	1	92.7	[97]
Rapeseed oil	KF/Eu_2O_3 (3.0)	12:1	65	1	92.5	[98]
	Cellulose derived sulfonated C (20.0)	20:1	65	6	93.4	[99]
Cottonseed oil	CH_3ONa (0.75)	6:1	65	1.5	96.9	[100]
	TiO_2/SO_4^{-2} (2.0)	12:1	230	8	>90	[101]
	KF-Al_2O_3 (3.0)	12:1	65	3	>90	[102]
Rice barn	CH_3ONa (0.88)	6:1	55	1	83.3	[103]
Castor oil	CH_3ONa (1.0)	16:1	30	0.5	93.1	[103]
	Ni-dopped ZnO nanoparticles (11.0)	8:1	60	1	95.20	[104]

(continued)

Table 2 (continued)

Feedstock	Catalyst (wt.%)	Alcohol to oil molar ratio	Temp. (°C)	Time (h)	Yield (%)	Refs.
	[b]SiO_2/50% H_2SO_4 (1.0)	6:1	n.a	0.5	95	[105]
	[b]SiO_2/50% H_2SO_4 (1.0)	6:1	n.a	25 min	95	[105]
	[b]Al_2O_3/50% KOH (1.0)	6:1	n.a	5 min	95	[105]
	[b]H_2SO_4/C (5.0)	12:1	65	1	94	[106]
Canola oil	$AlCl_3$ (5.0)	24:1	100	18	98	[107]
	[a]Dolomite (calcined) (5.3)	7.6:1	60	2.5	96.6	[108]
	Li/TiO_2 (5.0)	24:1	65	3	98	[109]
	[a]CaO (5.3)	7.5:1	60	2.5	99.4	[108]
Jatropha	CaO (1.5)	9:1	70	2.5	93	[110]
	CaMgO (4.0)	15:1	65	6	83	[111]
	KNO_3/Al_2O_3 (6.0)	12:1	70	6	84	[112]
	Sr^{2+}-CaO/MgO (5.0)	9:1	65	2	99.6	[113]
	[a]CaO (5.5)	11:1	64	n.a	95	[114]
	[a]Activated carbon supported heteropoly acid (20.0)	20:1	60	40 min	87.3	[115]
	[b]CaO (3.0)	10:1	65	6 min	78.1	[65]
	[b]Al_2O_3/CaO (3.0)	10:1	65	6 min	85.7	[65]
	[b]Fe_2O_3/CaO (3.0)	10:1	65	6 min	86.8	[65]
	[b]MnO_2/CaO (3.0)	10:1	65	6 min	93.4	[65]
	[b]ZnO/CaO (3.0)	10:1	65	6 min	99.1	[65]
	[b]CaO/P_2O_5 (4.0)	18:1	n.a	5 min	94	[116]
Karanja (*Pomgamia pinnata*)	Li/CaO (2.0)	12:1	65	8	94.8	[117]
	ZnO-methanol/C	10:1	105	1.5	–	[118]
	[b]CaO (3.0)	10:1	65	6 min	74.1	[65]
	[b]Al_2O_3/CaO (3.0)	10:1	65	6 min	85.4	[65]
	[b]Fe_2O_3/CaO (3.0)	10:1	65	6 min	87.1	[65]
	[b]MnO_2/CaO (3.0)	10:1	65	6 min	93	[65]
	[b]ZnO/CaO (3.0)	10:1	65	6 min	97.8	[65]
Algae	Sulfonated biochar (5.0)	10:1	100	1	98.2	[119]
	[b]CaO (3.0)	10:1	65	6 min	80.1	[65]
	[b]Al_2O_3/CaO (3.0)	10:1	65	6 min	86.3	[65]
	[b]Fe_2O_3/CaO (3.0)	10:1	65	6 min	88.2	[65]

(continued)

Table 2 (continued)

Feedstock	Catalyst (wt.%)	Alcohol to oil molar ratio	Temp. (°C)	Time (h)	Yield (%)	Refs.
	[b]MnO_2/CaO (3.0)	10:1	65	6 min	96.7	[65]
	[b]ZnO/CaO (3.0)	10:1	65	6 min	99.8	[65]
Waste cooking oil	Xylose derived sulfonated C catalyst (10.0)	10:1	110	4	89.6	[120]
	Zinc stearate immobilized on silica gel (3.0)	18:1	200	10	98.0	[121]
	[b]Sulfonated biochar-BS140 (10.0)	10:1	140	10 min	72	[122]

[a]Reaction carried out under ultrasonication; [b]Reactions carried out under microwave irradiation. n.a.: not available

technology is still in the developmental research stage. The challenges in identifying potential algal strains with satisfactory oil yield and insufficient information about their harvesting technology must be surmounted. Moreover, a detailed techno-economic analysis must be performed to find out the commercial feasibility of producing algal biodiesel. Active participation by the government with favorable policies and economic assistance (e.g., subsidy) is desired for the improvement and advancement of these technologies. Active collaboration and research activities should be promoted between the industries and academic institutions to explore advanced biodiesel technologies based on biorefineries and biotechnological approaches. The future of biodiesel will be impacted by the price of crude oil and the availability of alternative fuels.

2.2 *Hydrocarbon Fuels from Fatty Acids and Esters*

There is increasing interest in sourcing diesel-range, straight-chain, saturated hydrocarbons from fatty acids or their esters. Unlike biodiesel, liquid hydrocarbon fuels can be used directly as the drop-in replacement for petrodiesel. Fatty acids or esters can be reduced into hydrocarbons using overpressure of hydrogen and a suitable metal catalyst under hydrodeoxygenation (HDO) conditions [123, 124]. Triglycerides have also been used directly for HDO chemistry, forming saturated hydrocarbons in excellent yields [125]. In HDO, a temperature of around 180–250 °C and an H_2 overpressure of 1–4 MPa are typically used. Both noble metals and non-noble metals have shown good catalytic activity. Catalyst supports containing acidic sites were found to promote the hydrogenolysis of the C–O bond. Alternatively, the fatty acids or their esters can be subjected to catalytic decarboxylation (CDC) conditions

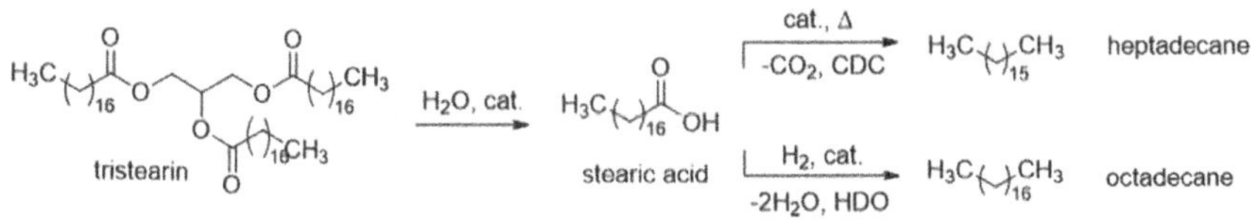

Scheme 2 Synthesis of hydrocarbon fuels from fatty acids and triglycerides

forming hydrocarbon fuels [126]. The CDC process expels oxygen atoms forming CO_2 without external hydrogen and is preferred for a better hydrogen economy (Scheme 2). Noble metals (e.g., Pd, Pt) and non-noble metals (e.g., Cu, Ni) have been employed as catalysts for the CDC reaction [127, 128]. Typically, a high reaction temperature (ca. 300–350 °C) and an acidic support (e.g., zeolite, hydrotalcite) for the metal catalysts promote the CDC process. High selectivity and yields of targeted hydrocarbons have been achieved in many cases. Analogous to biodiesel, the primary challenge here is to secure abundant and inexpensive sources of triglyceride.

2.3 Fuels from Glycerol

The enormous quantity of glycerol produced as the byproduct during biodiesel synthesis has drastically reduced its price and received significant attention as a renewable feedstock for biofuels. Both hydrocarbon fuels and fuel additives have been produced from glycerol. For example, glycerol has been converted into monocyclic aromatic hydrocarbons (e.g., BTX) over zeolite catalysts under pyrolytic conditions [129]. Glycerol can be selectively deoxygenated into propylene, which then can be oligomerized under heterogeneous acid catalysis forming naphtha and diesel-range hydrocarbons. A Fe-Mo bimetallic heterogeneous catalyst supported on carbon afforded around 67% yield of propylene under optimized conditions (300 °C, 8.0 MPa H_2, 6 h) [130]. Glycerol can be converted into various fuel oxygenates such as esters, acetal/ketal, and ethers by reacting with suitable carboxylic acids (e.g., acetic acid), ketones (e.g., acetone), and olefins (e.g., isobutylene). The fuel oxygenates from glycerol improve the octane rating and cold-flow properties of petrofuels and biofuels [131].

3 Fuels from Carbohydrate Fraction

The carbohydrate fraction (i.e., cellulose & hemicellulose) in lignocellulose is nearly 70% by mass [7]. In addition, the edible biomasses contain a significant quantity of sugars (e.g., sucrose, glucose, fructose). Even though the sugars and the polymeric carbohydrates can directly be converted into hydrocarbon-rich biofuels by pyrolysis,

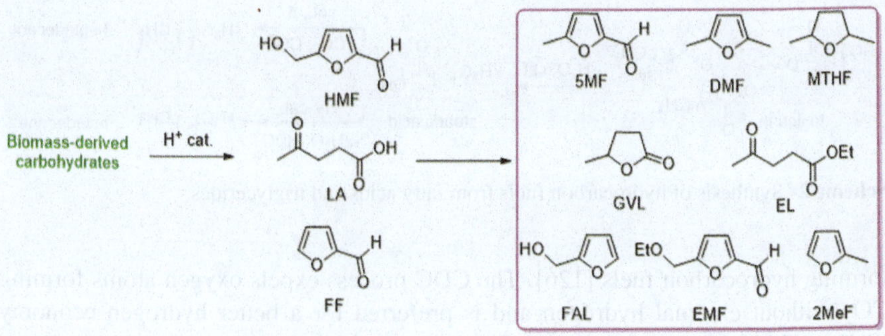

Scheme 3 Preparation of furanic and levulinic biofuels from the carbohydrate fraction of biomass

more selective processes are available. Glucose or cellulose may be reduced to n-hexane in a single step by the aqueous phase reforming process [132]. A more well-recognized route for preparing designed hydrocarbon fuels from carbohydrates is via the furanic and levulinic intermediates [133]. The acid hydrolysis of cellulose forms glucose, which subsequently undergoes a dehydration reaction forming 5-(hydroxymethyl)furfural (HMF) [134]. The century-old catalytic process reduces the oxygen content in the parent sugar by 50% without carbon–carbon bond-breaking reactions and forms water as the sole byproduct. The pentose sugars (e.g., xylose) in hemicellulose produce furfural (FF) by the acid-catalyzed hydrolysis and dehydration reactions [135]. HMF can get rehydrated and ring-opened to produce levulinic acid (LA) and formic acid in aqueous acid at elevated temperatures [136]. HMF, FF, LA, and their derivatives have been used as reactive chemical intermediates to build the carbon backbone of various fuels (Scheme 3). Recent advances in producing fuel oxygenates and hydrocarbon fuels from the carbohydrate fraction of biomass are discussed in the following subsections.

3.1 Fuel Oxygenates from Carbohydrates

As mentioned earlier, the fuel oxygenates are typically blended with hydrocarbon fuels for superior combustion characteristics and reduced emission. HMF, FF, or LA cannot be used directly as fuel additives. However, simple derivatization of the intermediates produces high-quality fuel oxygenates. For example, the esterification of LA with ethanol (renewably produced by fermentation of glucose/cellulose) produces ethyl levulinate (EL). EL can also be produced directly from HMF, furfuryl alcohol, carbohydrates, and raw biomasses. Like other synthetic methods, the most used catalysts include mineral acids, heteropoly acids and their hybrids, sulfonic-acid-based heterogeneous catalysts and resins, metal salts and supported metal oxides, and ionic liquids [137, 138]. EL has shown promise as a cleaner-burning diesel-additive. EL is a sulfur-free fuel with high lubricity, good anti-knocking index, high chemical stability, and acceptable cold-flow properties [139].

The furfurals (i.e., HMF, FF) are reduced under catalytic hydrogenation conditions to produce furanic compounds with lesser oxygen content, higher chemical stability, and increased energy density.

Selective hydrogenation of HMF or its analogs under catalytic conditions forms 2,5-dimethylfuran (DMF). DMF has a high research octane number (RON) of 119 and could be a good octane booster in gasoline fuels. DMF has an energy density of 33.7 MJ/kg, which is very similar to gasoline (46 MJ/kg) and much higher than ethanol (25 MJ/kg) [140, 141]. The detailed engine application studies of combustion and emission characteristics of DMF/gasoline blends have been studied [142]. DMF can also be used as a chemical intermediate for synthesizing *p*-xylene, which may be used as a high-octane aromatic fuel or as feedstock for polyethylene terephthalate. Hundreds of publications attempted the synthesis of DMF from HMF or directly from carbohydrates using a variety of catalysts and reaction conditions. Table 3 lists some recent literature on synthesizing DMF from HMF using efficient, tailor-made heterogeneous catalysts.

The use of noble [Table 3, entries 2 and 7], non-noble [Table 3, entries 4, 6 and 8], and bimetallic catalysts [Table 3, entries 1, 3 and 5] were reported for the transformation of HMF to DMF. The use of noble metal catalysts requires relatively milder reaction conditions than their non-noble metal counterparts. The synthesis of

Table 3 Representative catalysts for the synthesis of DMF from HMF

S/N	Catalyst	Reaction conditions	Yield (%)	Refs.
1	Cu-Co@C (8 wt.%)	180 °C, 8 h, ethanol, 5 MPa H_2	99.4	[143]
2	3% Pd/C (10 wt.%)	170 °C, 4 h, 2-proponol, 2.06 MPa H_2	99	[144]
3	Cu-Pd@C (20 wt.%)	120 °C, 7 h, THF, 1.5 MPa H_2	97	[145]
4	Ni-OMD (20 wt.%)	200 °C, 6 h, water, 3.0 MPa H_2	98.7	[146]
5	$Pd_{50}Au_{50}/C$ (25 wt.%)	60 °C, 12 h, THF, H_2 balloon (purge), HCl	99.9	[147]
6	$LaFeO_3/Ni20$ (79 wt.%)	230 °C, 6 h, ethanol, 5.0 MPa H_2	98.3	[148]
7	Ru/ZSM-5 (50 wt.%)	180 °C, 3 h, ethanol, 1.7 MPa H_2	95	[149]
8	NC-Cu/MgAlO (50 wt.%)	220 °C, 0.5 h, cyclohexanol	96	[150]
9[a]	CuCo@C-TsOH-1M (100 wt.%)	220 °C, 10 h, THF, 3 MPa	71.1	[151]

Abbreviations: OMD: Organic method deposition method, TsOH: p-Toluenesulfonic acid, THF: Tetrahydrofuran. [a]Fructose as a starting material

DMF from HMF requires a series of hydrogenation and hydrogenolysis steps without damaging the furan ring. Detailed analysis of the catalysts has been carried out for selective, high-yielding synthesis of DMF using inexpensive, efficient, and highly recyclable catalysts. Attempts to produce DMF directly from carbohydrates are also not uncommon in the literature [Table 1, entry 9]. DMF can also be synthesized from structural analogs of HMF in the liquid phase [152]. Before the commercial adoption of DMF as a biofuel, the challenges to solve are ensuring the availability of HMF from inexpensive feedstock in satisfactory yield and developing an efficient catalytic system that functions under energy-economic conditions.

2,5-Dimethyltetrahydrofuran (DMTHF) can be prepared by the catalytic reduction of HMF or directly from carbohydrates by concomitant dehydration and hydrogenation reactions using multi-tasking catalysts. DMTHF has a high energy density (32.8 MJ/kg), decent volatility (b.p. 90 °C), high octane number (RON = 82), and low water miscibility. Chen et al. reported the synthesis of DMTHF in a 97% yield (180 °C, 3 MPa H_2, 10 h) from HMF using a non-promoted Ni/SBA-15 catalyst [153]. The synthesis of DMTHF directly from biomass-derived hexose sugars was reported using dual catalytic systems [154, 155].

HMF or its hydrophobic analogs can be transformed into 5-(ethoxymethyl)furfural (EMF) by etherification. EMF can also be produced from various hydrophobic analogs of HMF such as 5-(chloromethyl)furfural and directly from carbohydrates or cellulosic biomass by acid-catalyzed ethanolysis [156]. EMF possesses superior fuel properties, such as high energy density and boiling point, low toxicity, high cetane number, low soot emission, and excellent combustion performance.

Catalytic hydrogenation of the aldehyde group in FF forms 2-methylfuran (2MeF), a promising biofuel. Pd, Pt, Ru, and Ni-based noble and Cu, Co, and Fe-based non-noble metal heterogeneous catalysts were reported in the literature. At the same time, the use of bimetallic catalysts showed superior catalytic activities compared to their respective metals. The alcoholic solvents can be used as hydrogen donors to achieve a good yield of 2MeF [157].

The ring-hydrogenation of 2MeF leads to 2-methyltetrahydrofuran (MTHF), a potential fuel oxygenate. One-pot synthesis of MTHF was reported from FF [158]. Upare et al. reported the synthesis of MTHF from LA in a 64% yield (265 °C, 2.5 MPa H_2) using a copper/silica nanocomposite catalyst [159]. MTHF has been blended up to 70% with gasoline [160].

The ketone group in LA or its esters can be reduced into alcohol, which immediately undergoes intramolecular lactonization forming γ-valerolactone (GVL) [161, 162]. Comprehensive reviews on the targeted synthesis of GVL from LA and studies on fuel properties have been published in the literature. Even though non-noble metal catalysts have been used and shown promising results, noble metals such as ruthenium are routinely employed as a catalyst for the hydrogenation reaction. GVL can be directly used as a fuel additive in diesel fuels. The major issue associated with GVL as a fuel additive is its solubility in water. GVL can be further hydrogenated into MTHF and valeric esters.

3.2 Hydrocarbon Fuels from Carbohydrates

Aviation fuels (or jet-fuels) are in the C_{10}–C_{15} range, whereas diesel is in the C_{15}–C_{22} range straight-chain or minimally branched linear or cyclic compounds. On the other hand, gasoline is C_6–C_{10} saturated branched hydrocarbons and monocyclic aromatic hydrocarbons (e.g., toluene). The strategy routinely adopted for the targeted synthesis of hydrocarbon fuels from the carbohydrate fraction is to initially produce some reactive chemical intermediates, which are then coupled together to extend the carbon chain in the desired range. Finally, the oxygenated compounds are subjected to HDO chemistry under catalytic conditions to produce hydrocarbons (Fig. 3) [133].

Interestingly, the sugar-derived chemical intermediates (e.g., furfurals, levulinic acid, sorbitol) contain a straight-chain carbon backbone. The carbon backbone in HMF (a C_6 compound) and FF (a C_5 compound) intermediates are then extended using various C–C bond forming reactions such as aldol condensation, benzoin condensation, Friedel–Crafts reaction, Diels–Alder reaction, and Michael addition [163]. The coupling partners are also preferentially bio-derived molecules. Biogenic ketones (e.g., acetone), other renewable ketones (e.g., cyclopentanone), and various furanic and levulinic products have been used as coupling partners (Scheme 4) [164, 165]. In the final reduction step, hydrogen (molecular hydrogen or a hydrogen donor solvent) is used as the reagent to eliminate oxygen atoms from the oxygenated compounds by HDO and CDC conditions, where water and carbon dioxide are formed as the byproduct, respectively. The HDO process allows retaining all the

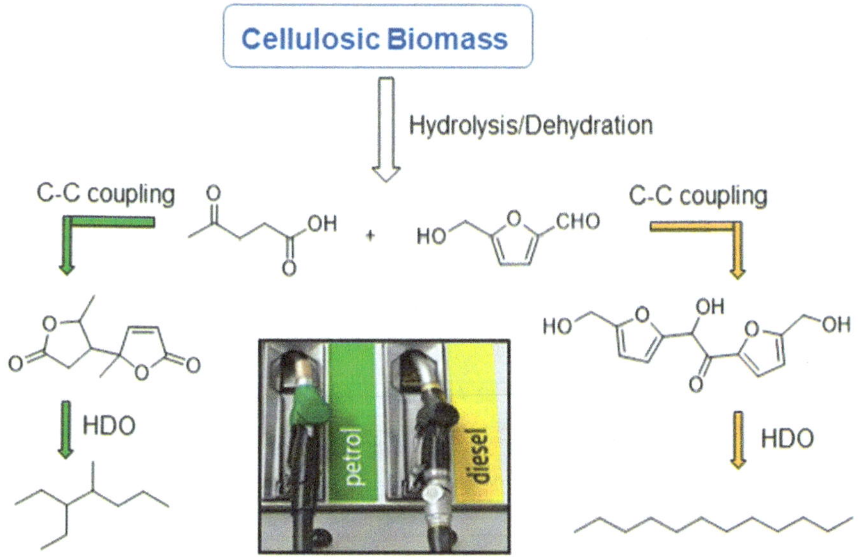

Fig. 3 The synthesis of gasoline- and diesel-range hydrocarbon fuels from cellulosic biomass via furanic and levulinic intermediates

carbon atoms in the feedstock and forms water as the sole byproduct. However, hydrogen gas is typically used in high overpressure, and more equivalent of H_2 is required in the HDO process than in the CDC process. Since the carbohydrate-derived chemical intermediates and the biogenic ketones have mostly liner structures, the oxygenates produced from them also have linear or lightly substituted structures. For example, lignocellulose-derived 2MeF, butanal, and acetone have been condensed into diesel-range hydrocarbon fuels by HDO chemistry [166]. Therefore, the synthesis of aviation- or diesel-range fuels from the carbohydrate-derived molecules is well-studied in the literature.

The targeted synthesis of heavily branched gasoline-range fuels directly from carbohydrate-derived chemical intermediates was first reported by Mascal et al. in 2014 [167]. The process started with biomass-derived LA. In the first step, angelica lactones (ALs) were produced by the intramolecular lactonization of LA catalyzed by a montmorillonite clay (K10) catalyst. ALs were then conveniently dimerized into angelica lactone dimer (ALD) under base catalysis. Finally, ALD was subjected to the HDO chemistry using an Ir-ReO_x catalyst producing C_6–C_{10} (gasoline-range) branched hydrocarbons in 78% yield (Scheme 5A). In a later report, the same research group reported the hydrogen-economic synthesis of branched linear and cyclic hydrocarbons in nearly quantitative yield by the CDC of ALD using a Pd/C catalyst [168]. The novel approaches encouraged other research groups to develop many different pathways over the past few years.

In related work, EL was self-condensed using NaOEt as a base catalyst, and the cyclic C_{10} oxygenate was then subjected to catalytic hydrogenation conditions in the presence of a Pd-Al_2O_3 catalyst under hydrogen overpressure (Scheme 5B) [169]. The combination of HDO and CDC processes afforded a C_8–C_{10} substituted cyclopentane.

Two molecules of LA were coupled by Kolbe electrolysis to form 2,7-octanedione (ODO) (Scheme 5C). ODO was then converted into cyclic ketones by intramolecular cyclization and hydrogenation reactions using p-toluenesulfonic acid and Pd/C as catalysts under a hydrogen atmosphere. In the final step, branched cyclic hydrocarbons were produced using Pd/C as the hydrogenation catalyst and $Al(OTf)_3$ as a Lewis acid catalyst for promoting the HDO process [170]. As mentioned earlier, the

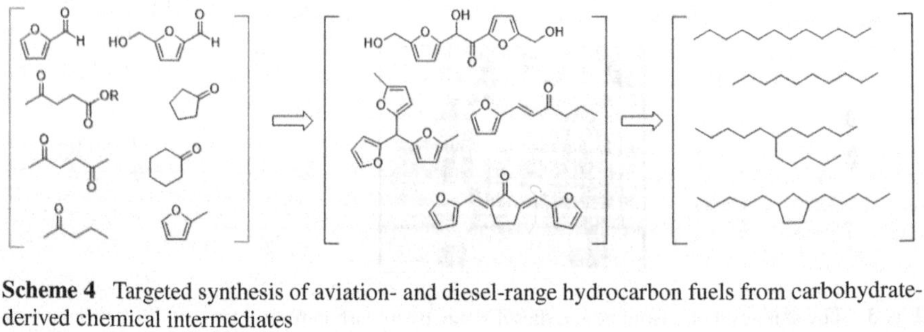

Scheme 4 Targeted synthesis of aviation- and diesel-range hydrocarbon fuels from carbohydrate-derived chemical intermediates

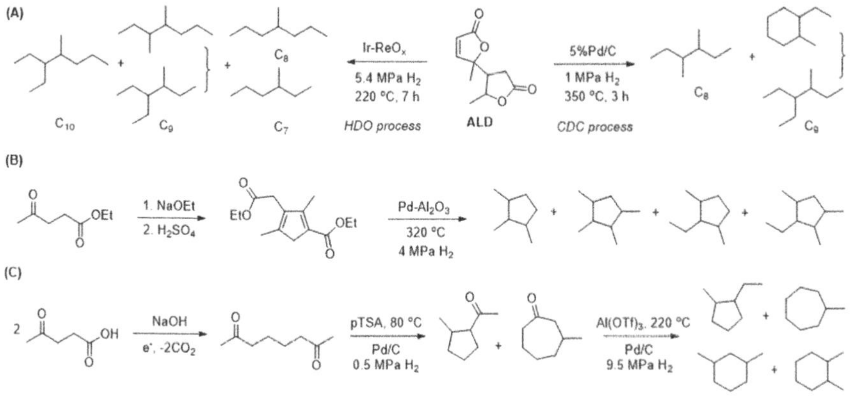

Scheme 5 The preparation of gasoline-like hydrocarbons from ALD under catalytic conditions

primary advantage of these fuels is the molecular control over structure and properties. The challenges include multi-step synthesis and separation (also purification) of the chemical intermediates. Even though excellent overall yields of hydrocarbon fuels have been achieved in several cases, attention must be given to the commercial viability of producing the furanic and levulinic intermediates required for their synthesis. Moreover, external hydrogen and expensive noble metal catalysts must be curtailed for the processes.

4 Fuels from Lignin

Lignin is the second most abundant biopolymer after cellulose and is produced in large quantities in the paper processing industry during the pulping process [171]. This unique polymer comprises aromatic units connected by methylene, ether, and carbonyl groups [172]. Various types of lignin, such as kraft lignin, lignosulfonate lignin, organosolv lignin, and soda lignin, are known [173]. In addition, lignin is produced as a byproduct after hydrolyzing the carbohydrate fraction of lignocellulosic biomass under acid catalysis. Due to its high calorific value and low oxygen content, the lignin fraction is typically separated and combusted to generate the process heat required for other biorefinery operations, such as cellulose-centric industries [174]. Alternatively, lignin can be pyrolyzed into bio-oil or gasified into syngas for further valorization (discussed in the next section). Recently, there have been significant interests in the targeted synthesis of aromatic biofuels from lignin [175, 176]. Interestingly, many processes attempt the value addition strategy on lignin model compounds (e.g., *p*-coumaryl alcohol, coniferyl alcohol, sinapyl alcohol) and claim that the process may also be extended to native or modified lignin (Scheme 6) [177]. The primary challenge for converting lignin into aromatic hydrocarbons is to control the extensive C–C and C–O cleavage reactions [178]. Pyrolysis of lignin

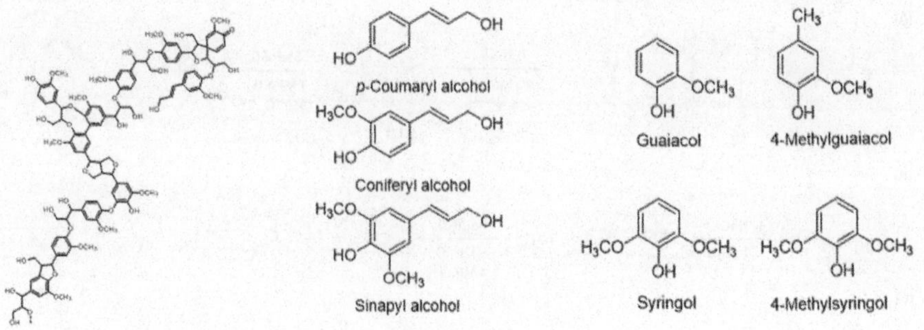

Scheme 6 The polymeric structure of native softwood lignin (*left*), lignin model compounds (*middle*), and phenolic intermediates obtained from lignin (*right*)

initially cleaves the inter-unit linkages forming intermediates such as guaiacol and syringol, which then transform into aromatic hydrocarbons by dehydroxylation and demethoxylation reactions [179]. The monocyclic aromatic hydrocarbons (benzene, toluene, ethylbenzene, and xylenes) could be used as high-octane fuels and as feedstock for synthesizing chemicals.

The composition of pyrolysis oil derived from lignin depends on the type of lignin used, reaction parameters, and the catalyst employed. Thermally stable zeolite-based catalysts have shown promising to convert lignin into hydrocarbon-rich biofuels [180]. Lignin can also be depolymerized and transformed into monocyclic hydrocarbons by the HDO process using molecular hydrogen in the presence of a suitable metal catalyst (e.g., Ru). Lignin was depolymerized under transfer hydrogenation conditions using Raney-Ni as the catalyst and 2-propanol as the hydrogen donor in a novel strategy [181]. The resulting oily product containing mostly phenolic species and their oligomers was then subjected to HDO chemistry over phosphidated Ni/SiO_2 to give aromatic or aliphatic hydrocarbons in the gasoline range. Interestingly, the relative ratio of aromatic and aliphatic compounds in the product could be controlled by simply manipulating the H_2 overpressure, with lower H_2 pressure favoring the aromatic hydrocarbon formation.

5 Pyrolysis and Gasification of Biomass

Pyrolysis and gasification of biomass are fast (seconds to minutes), feedstock independent, and benefitted from matured technology [182]. Biomass pyrolysis produces solid biochar, pyrolytic oil (or bio-oil), and combustible gases. Bio-oil is a complex mixture of compounds bearing diverse structures and functionalities, which often mandates synthetic upgrading (e.g., hydrogenation, fractionation) before use in boilers and engines [183]. The use of catalysts in biomass pyrolysis helps lower the severity of reaction conditions and improve selectivity toward desired molecules, i.e., hydrocarbons. Catalysts help reduce the amount of oxygenates in pyrolysis oil or bio-oil, thereby improving its quality (e.g., high calorific value, enhanced stability). The chemical composition and yield of bio-oil depend on several parameters, including the biomass used, reaction temperature, type of catalyst, residence time, the particle size of feedstock, and heating rate. Temperature around 500–550 °C and a high heating rate are typically used for the highest bio-oil yield [184]. Aluminosilicates and zeolites are frequently employed to produce bio-oil rich in aromatic hydrocarbons with superior combustion properties. The acidic sites on the catalysts promote C–O bond cleavage reactions reducing the oxygen content in the product molecules forming water as an innocuous byproduct. Catalysts containing basic sites lead to more gaseous products rich in H_2 [185]. The biomass gasification is also performed in the presence of a suitable catalyst at temperatures higher than pyrolysis temperature (ca. 500–750 °C), forming a gaseous mixture rich in CO and H_2 [186]. The undesired gases (e.g., CO_2) are eliminated in the next step. The molar ratio of CO and H_2 is then adjusted and subjected to gas-to-liquid technologies, such as the Fischer-Tropsch (F-T) process for making linear hydrocarbons [187]. Silica and silica-alumina catalysts exchanged with suitable metal ions (e.g., K^+) are commonly employed as catalysts that help lower the temperature required for gasification and higher selectivity toward H_2 [188]. Various metal oxides, such as CaO, TiO_2, Al_2O_3, and ZrO_2, were used as the catalyst for steam gasification of biomass into hydrogen-rich syngas [189]. The F-T process is an established technology that converts the syngas into liquid hydrocarbons by a series of catalytic steps. The hydrocarbon mixture is then fractionated into gasoline and diesel-range fuels. The major drawback of these technologies is their high energy demand. Although some level of selectivity is achieved by using catalysts, it is significantly lower than other catalytic processes discussed in this chapter. A mixture of compounds is inevitably formed in the above cases, which must be separated or upgraded before using as transportation fuels.

6 Conclusion and Future Directions

In summary, there have been significant research breakthroughs over the past decades on the designed synthesis of renewable liquid transportation fuels by the catalytic transformation of various biomass components. The choice of biomass for biofuels synthesis remains paramount from the availability, cost, environmental impact, and conversion technology perspectives. The target is to select biomass feedstock that is carbon-neutral or preferably carbon-negative for maximizing the ecological incentives. A detailed life cycle assessment of the biomass feedstock considering the materials and energy input during its production is warranted before adopting it in a biorefinery setting for biofuels synthesis. Heterogeneous catalysis remains at the heart of the chemical-catalytic value addition of biomass into biofuels. Tailor-made catalysts are being developed that show high catalytic activity under mild reaction conditions, high selectivity and yield targeted product(s), high stability, and good recyclability. Detailed mechanistic studies (theoretical and experimental) of biofuels synthesis and advanced characterization techniques for heterogeneous catalysts have helped design more efficient catalysts. In this regard, Various C–C bond-forming reactions are intelligently used for building the carbon framework of aviation-, gasoline-, and diesel-range fuels starting from biomass-derived chemical intermediates, such as furanics and levulinates. Molecular hydrogen is crucial for synthesizing or upgrading various types of biofuels. Even though green hydrogen by electrolysis of water is an upcoming technology, biohydrogen from biomass via gasification or aqueous reforming strategy could prove to be vital for a circular hydrogen economy and biofuels production. Finally, it may be said that apart from the affordability and technological preparedness, the commercial prospect of biofuels production will be shaped by government policies, environmental regulations, and public outlook for a sustainable future.

References

1. Stevens P (2018) The role of oil and gas in the economic development of the global economy. Extr Ind. https://doi.org/10.1093/oso/9780198817369.003.0004
2. Escobar JC, Lora ES, Venturini OJ, Yáñez EE, Castillo EF, Almazan O (2009) Biofuels: environment, technology and food security. Renew Sustain Energy Rev 13:1275–1287
3. https://www.eia.gov/tools/faqs/faq.php
4. Mitchell JV, Stevens P (2009) Ending dependence: hard choices for oil. Chatham House.
5. Kheshgi HS, Prince RC, Marland G (2000) The potential of biomass fuels in the context of global climate change: focus on transportation fuels. Annu Rev Energy Environ 25:199–244
6. Matthews LR, Niziolek AM, Onel O, Pinnaduwage N, Floudas CA (2016) Biomass to liquid transportation fuels via biological and thermochemical conversion: process synthesis and global optimization strategies. Ind Eng Chem Res 55:3203–3225
7. Vassilev SV, Baxter D, Andersen LK, Vassileva CG (2010) An overview of the chemical composition of biomass. Fuel 89:913–933
8. Rogers KA, Zheng Y (2016) Selective deoxygenation of biomass-derived bio-oils within hydrogen-modest environments: a review and new insights. ChemSuschem 9:1750–1772
9. Gasparatos A, Stromberg P, Takeuchi K (2013) Sustainability impacts of first-generation biofuels. Anim Front 3:12–26
10. Ganguly P, Sarkhel R, Das P (2021) The second- and third-generation biofuel technologies: comparative perspectives. In: Dutta S, Mustansar Hussain C (eds) Sustainable fuel technologies handbook. Academic Press, pp 29–50
11. Haq I, Qaisar K, Nawaz A, Akram F, Mukhtar H, Zhou X, Xu Y, Mumtaz MW, Rashid U, Ghani WAWAK, Chong TSY (2021) Advances in valorization of lignocellulosic biomass towards energy generation. Catalysts 11:309
12. Jahirul MI, Rasul MG, Chowdhury AA, Ashwath N (2012) Biofuels production through biomass pyrolysis—a technological review. Energies 5:4952–5001
13. Ail SS, Dasappa S (2016) Biomass to liquid transportation fuel via Fischer Tropsch synthesis–Technology review and current scenario. Renew Sustain Energy Rev 58:267–286
14. Binod P, Gnansounou E, Sindhu R, Pandey A (2019) Enzymes for second generation biofuels: recent developments and future perspectives. Bioresour Technol Rep 5:317–325
15. Zhou C-H, Xia X, Lin C-X, Tong D-S, Beltramini J (2011) Catalytic conversion of lignocellulosic biomass to fine chemicals and fuels. Chem Soc Rev 40:5588–5617
16. Anastas PT, Kirchhoff MM, Williamson TC (2001) Catalysis as a foundational pillar of green chemistry. Appl Catal A 221:3–13

17. Di Serio M, Tesser R, Pengmei L, Santacesaria E (2008) Heterogeneous catalysts for biodiesel production. Energy Fuels 22:207–217
18. Lu Y, Zhang Z, Wang H, Wang Y (2021) Toward efficient single-atom catalysts for renewable fuels and chemicals production from biomass and CO_2. Appl Catal B 292:120162
19. Akia M, Yazdani F, Motaee E, Han D, Arandiyan H (2014) A review on conversion of biomass to biofuel by nanocatalysts. Biofuel Res J 1:16–25
20. Karatzos S, van Dyk JS, McMillan JD, Saddler J (2017) Drop-in biofuel production via conventional (lipid/fatty acid) and advanced (biomass) routes. Part I. Biofuels, Bioprod Bioref 11:344–362
21. McCormick RL, Ratcliff MA, Christensen E, Fouts L, Luecke J, Chupka GM, Yanowitz J, Tian M, Boot M (2015) Properties of oxygenates found in upgraded biomass pyrolysis oil as components of spark and compression ignition engine fuels. Energy Fuels 29:2453–2461
22. Díaz-Pérez MA, Serrano-Ruiz JC (2020) Catalytic production of jet fuels from biomass. Molecules 25:802
23. Tao BY (2007) Industrial applications for plant oils and lipids. In: Yang S-T (ed) Bioprocessing for value-added products from renewable resources. Elsevier, Amsterdam, pp 611–627
24. Orsavova J, Misurcova L, Ambrozova JV, Vicha R, Mlcek J (2015) Fatty acids composition of vegetable oils and its contribution to dietary energy intake and dependence of cardiovascular mortality on dietary intake of fatty acids. Int J Mol Sci 16:12871–12890
25. Enweremadu CC, Mbarawa MM (2009) Technical aspects of production and analysis of biodiesel from used cooking oil—a review. Renew Sustain Energy Rev 13:2205–2224
26. Singh SP, Singh D (2010) Biodiesel production through the use of different sources and characterization of oils and their esters as the substitute of diesel: a review. Renew Sustain Energy Rev 14:200–216
27. Joshi G, Pandey JK, Rana S, Rawat DS (2017) Challenges and opportunities for the application of biofuel. Renew Sustain Energy Rev 79:850–866
28. Ambat I, Srivastava V, Sillanpää M (2018) Recent advancement in biodiesel production methodologies using various feedstock: a review. Renew Sustain Energy Rev 90:356–369
29. Singh D, Sharma D, Soni SL, Sharma S, Kumar Sharma P, Jhalani A (2020) A review on feedstocks, production processes, and yield for different generations of biodiesel. Fuel 262:116553
30. Gad MS, Ismail MA (2021) Effect of waste cooking oil biodiesel blending with gasoline and kerosene on diesel engine performance, emissions and combustion characteristics. Process Saf Environ Prot 149:1–10
31. Xiao H, Guo F, Wang R, Yang X, Li S, Ruan J (2020) Combustion performance and emission characteristics of diesel engine fueled with iso-butanol/biodiesel blends. Fuel 268:117387
32. Marchetti JM, Miguel VU, Errazu AF (2007) Possible methods for biodiesel production. Renew Sustain Energy Rev 11:1300–1311
33. Abbaszaadeh A, Ghobadian B, Omidkhah MR, Najafi G (2012) Current biodiesel production technologies: a comparative review. Energy Convers Manage 63:138–148
34. Athar M, Zaidi S (2020) A review of the feedstocks, catalysts, and intensification techniques for sustainable biodiesel production. J Environ Chem Eng 8:104523
35. Bhuiya MMK, Rasul MG, Khan MMK, Ashwath N, Azad AK (2016) Prospects of 2nd generation biodiesel as a sustainable fuel—Part: 1 selection of feedstocks, oil extraction techniques and conversion technologies. Renew Sustain Energy Rev 55:1109–1128
36. Fasanya OO, Osigbesan AA, Avbenake OP (2021) Biodiesel production from non-edible and waste lipid sources. In: Biodiesel technology and applications. Wiley, pp 389–427
37. Kim D-S, Hanifzadeh M, Kumar A (2018) Trend of biodiesel feedstock and its impact on biodiesel emission characteristics. Environ Prog Sustain Energy 37:7–19
38. Milledge JJ, Heaven S (2014) Methods of energy extraction from microalgal biomass: a review. Rev Environ Sci Biotechnol 13:301–320
39. Roick C, Otun KO, Diankanua N, Joshua G (2021) Non-edible feedstock for biodiesel production. In: Biodiesel technology and applications. Wiley, pp 285–309

40. Shokravi H, Shokravi Z, Heidarrezaei M, Ong HC, Rahimian Koloor SS, Petrů M, Lau WJ, Ismail AF (2021) Fourth generation biofuel from genetically modified algal biomass: challenges and future directions. Chemosphere 285:131535
41. Moser BR (2012) Preparation of fatty acid methyl esters from hazelnut, high-oleic peanut and walnut oils and evaluation as biodiesel. Fuel 92:231–238
42. Anjum SS, Prakash O, Pal A (2019) Conversion of non-edible Argemone Mexicana seed oil into biodiesel through the transesterification process. Energy Sources Part A 41:2356–2363
43. Rahman WU, Fatima A, Anwer AH, Athar M, Khan MZ, Khan NA, Halder G (2019) Biodiesel synthesis from eucalyptus oil by utilizing waste egg shell derived calcium based metal oxide catalyst. Process Saf Environ Prot 122:313–319
44. Awad S, Loubar K, Tazerout M (2014) Experimental investigation on the combustion, performance and pollutant emissions of biodiesel from animal fat residues on a direct injection diesel engine. Energy 69:826–836
45. Mata TM, Cardoso N, Ornelas M, Neves S, Caetano NS (2011) Evaluation of two purification methods of biodiesel from beef tallow, pork lard, and chicken fat. Energy Fuels 25:4756–4762
46. Sahar SS, Iqbal J, Ullah I, Bhatti HN, Nouren S, Habib-ur-Rehman NJ, Iqbal M (2018) Biodiesel production from waste cooking oil: an efficient technique to convert waste into biodiesel. Sustain Cities Soc 41:220–226
47. Dhawane SH, Karmakar B, Ghosh S, Halder G (2018) Parametric optimisation of biodiesel synthesis from waste cooking oil via Taguchi approach. J Environ Chem Eng 6:3971–3980
48. Yuliana M, Santoso SP, Soetaredjo FE, Ismadji S, Angkawijaya AE, Irawaty W, Ju Y-H, Tran-Nguyen PL, Hartono SB (2020) Utilization of waste capiz shell–based catalyst for the conversion of leather tanning waste into biodiesel. J Environ Chem Eng 8:104012
49. Abdullah B, Syed Muhammad SAF, Shokravi Z, Ismail S, Kassim KA, Mahmood AN, Aziz MMA (2019) Fourth generation biofuel: a review on risks and mitigation strategies. Renew Sustain Energy Rev 107:37–50
50. Rohith Renish R, Amala Justus Selvam M (2021) A critical review on production process, physicochemical properties, performance and emission characteristics of sea mango biodiesel-diesel blends. Mater Today: Proceedings 44:2600–2605
51. Godbole V, Pal MK, Gautam P (2021) A critical perspective on the scope of interdisciplinary approaches used in fourth-generation biofuel production. Algal Res 58:102436
52. Callegari A, Bolognesi S, Cecconet D, Capodaglio AG (2020) Production technologies, current role, and future prospects of biofuels feedstocks: a state-of-the-art review. Crit Rev Environ Sci Technol 50:384–436
53. Yaşar F (2020) Comparision of fuel properties of biodiesel fuels produced from different oils to determine the most suitable feedstock type. Fuel 264:116817
54. Naik SN, Goud VV, Rout PK, Dalai AK (2010) Production of first and second generation biofuels: a comprehensive review. Renew Sustain Energy Rev 14:578–597
55. Kumbhar MB, Lokhande PE, Chavan US, Salunkhe VG (2021) A global scenario of sustainable technologies and progress in a biodiesel production. In: Biodiesel technology and applications. Wiley, pp 215–240
56. Kulkarni MG, Dalai AK (2006) Waste cooking oil an economical source for biodiesel: a review. Ind Eng Chem Res 45:2901–2913
57. Yaakob Z, Mohammad M, Alherbawi M, Alam Z, Sopian K (2013) Overview of the production of biodiesel from waste cooking oil. Renew Sustain Energy Rev 18:184–193
58. Adewale P, Dumont M-J, Ngadi M (2015) Recent trends of biodiesel production from animal fat wastes and associated production techniques. Renew Sustain Energy Rev 45:574–588
59. Shields-Menard SA, Amirsadeghi M, French WT, Boopathy R (2018) A review on microbial lipids as a potential biofuel. Bioresour Technol 259:451–460
60. Azad AK, Rasul MG, Khan MMK, Sharma SC, Hazrat MA (2015) Prospect of biofuels as an alternative transport fuel in Australia. Renew Sustain Energy Rev 43:331–351
61. Aransiola EF, Ojumu TV, Oyekola OO, Madzimbamuto TF, Ikhu-Omoregbe DIO (2014) A review of current technology for biodiesel production: state of the art. Biomass Bioenergy 61:276–297

62. Renzaho AMN, Kamara JK, Toole M (2017) Biofuel production and its impact on food security in low and middle income countries: implications for the post-2015 sustainable development goals. Renew Sustain Energy Rev 78:503–516
63. Gerbens-Leenes PW (2017) Bioenergy water footprints, comparing first, second and third generation feedstocks for bioenergy supply in 2040. European Water 373–380.
64. Rawat DS, Joshi G, Lamba BY, Tiwari AK, Mallick S (2014) Impact of additives on storage stability of Karanja (*Pongamia Pinnata*) biodiesel blends with conventional diesel sold at retail outlets. Fuel 120:30–37
65. Joshi G, Rawat DS, Sharma AK, Pandey JK (2016) Microwave enhanced alcoholysis of non-edible (algal, jatropha and pongamia) oils using chemically activated egg shell derived CaO as heterogeneous catalyst. Bioresour Technol 219:487–492
66. Alalwan HA, Alminshid AH, Aljaafari HAS (2019) Promising evolution of biofuel generations. Subject review. Renew Energy Focus 28:127–139.
67. Mathimani T, Baldinelli A, Rajendran K, Prabakar D, Matheswaran M, Pieter van Leeuwen R, Pugazhendhi A (2019) Review on cultivation and thermochemical conversion of microalgae to fuels and chemicals: process evaluation and knowledge gaps. J Cleaner Prod 208:1053–1064
68. Mathimani T, Pugazhendhi A (2019) Utilization of algae for biofuel, bio-products and bio-remediation. Biocatal Agric Biotechnol 17:326–330
69. Chi NTL, Duc PA, Mathimani T, Pugazhendhi A (2019) Evaluating the potential of green alga Chlorella sp. for high biomass and lipid production in biodiesel viewpoint. Biocatal Agric Biotechnol 17:184–188
70. Lakaniemi A-M, Tuovinen OH, Puhakka JA (2013) Anaerobic conversion of microalgal biomass to sustainable energy carriers—a review. Bioresource Technol 135:222–231
71. Mat Aron NS, Khoo KS, Chew KW, Show PL, Chen W-H, Nguyen THP (2020) Sustainability of the four generations of biofuels—a review. Int J Energy Res 44:9266–9282
72. Atabani AE, Silitonga AS, Badruddin IA, Mahlia TMI, Masjuki HH, Mekhilef S (2012) A comprehensive review on biodiesel as an alternative energy resource and its characteristics. Renew Sustain Energy Rev 16:2070–2093
73. Jain P (2021) Biodiesel-processing, economics & potential in India. J Prod Oper Manag Econ (JPOME) 1:23–30
74. Gebremariam SN, Marchetti JM (2018) Economics of biodiesel production: review. Energy Convers Manage 168:74–84
75. Rodrigues A, Bordado JC, Santos RG (2017) Upgrading the glycerol from biodiesel production as a source of energy carriers and chemicals—a technological review for three chemical pathways. Energies 10:1817
76. Li X, Xiong Q, Wang M, Huang H, Yue G, Jin M (2021) In-situ corn fiber conversion method unlocks the role of viscosity on enhancing ethanol yield by reducing side-product glycerol. Ind Crops Prod 169:113653
77. Meher LC, Vidya Sagar D, Naik SN (2006) Technical aspects of biodiesel production by transesterification—a review. Renew Sustain Energy Rev 10:248–268
78. Reyero I, Arzamendi G, Gandía LM (2014) Heterogenization of the biodiesel synthesis catalysis: CaO and novel calcium compounds as transesterification catalysts. Chem Eng Res Des 92:1519–1530
79. Bet-Moushoul E, Farhadi K, Mansourpanah Y, Nikbakht AM, Molaei R, Forough M (2016) Application of CaO-based/Au nanoparticles as heterogeneous nanocatalysts in biodiesel production. Fuel 164:119–127
80. Babu NS, Sree R, Prasad PSS, Lingaiah N (2008) Room-temperature transesterification of edible and nonedible oils using a heterogeneous strong basic Mg/La catalyst. Energy Fuels 22:1965–1971
81. Liu X, He H, Wang Y, Zhu S, Piao X (2008) Transesterification of soybean oil to biodiesel using CaO as a solid base catalyst. Fuel 87:216–221
82. Fernandes FAN, Lopes RM, Mercado MP, Siqueira ES (2016) Production of soybean ethanol-based biodiesel using CaO heterogeneous catalysts promoted by Zn, K and Mg. Int J Green Energy 13:417–423

83. Liu X, Piao X, Wang Y, Zhu S (2008) Calcium ethoxide as a solid base catalyst for the transesterification of soybean oil to biodiesel. Energy Fuels 22:1313–1317
84. Li X, Lu G, Guo Y, Guo Y, Wang Y, Zhang Z, Liu X, Wang Y (2007) A novel solid superbase of Eu_2O_3/Al_2O_3 and its catalytic performance for the transesterification of soybean oil to biodiesel. Catalysis Commun 8:1969–1972
85. Xie W, Li H (2006) Alumina-supported potassium iodide as a heterogeneous catalyst for biodiesel production from soybean oil. J Mol Catal A: Chem 255:1–9
86. Xie W, Peng H, Chen L (2006) Transesterification of soybean oil catalyzed by potassium loaded on alumina as a solid-base catalyst. Appl Catal A 300:67–74
87. Garcia CM, Teixeira S, Marciniuk LL, Schuchardt U (2008) Transesterification of soybean oil catalyzed by sulfated zirconia. Bioresource Technol 99:6608–6613
88. Xie W, Huang X, Li H (2007) Soybean oil methyl esters preparation using NaX zeolites loaded with KOH as a heterogeneous catalyst. Bioresource Technol 98:936–939
89. Torres-Rodríguez DA, Romero-Ibarra IC, Ibarra IA, Pfeiffer H (2016) Biodiesel production from soybean and Jatropha oils using cesium impregnated sodium zirconate as a heterogeneous base catalyst. Renew Energy 93:323–331
90. Xie W, Peng H, Chen L (2006) Calcined Mg–Al hydrotalcites as solid base catalysts for methanolysis of soybean oil. J Mol Catal A: Chem 246:24–32
91. Ye X, Fernando S, Wilson W, Singh A (2007) Application of amphiphilic catalysts, ultrasonication, and nanoemulsions for biodiesel production process. Chem Eng Technol 30:1481–1487
92. Cintas P, Mantegna S, Gaudino EC, Cravotto G (2010) A new pilot flow reactor for high-intensity ultrasound irradiation. Application to the synthesis of biodiesel. Ultrason Sonochem 17:985–989
93. Choudhury HA, Chakma S, Moholkar VS (2014) Mechanistic insight into sonochemical biodiesel synthesis using heterogeneous base catalyst. Ultrason Sonochem 21:169–181
94. Hsiao M-C, Lin C-C, Chang Y-H (2011) Microwave irradiation-assisted transesterification of soybean oil to biodiesel catalyzed by nanopowder calcium oxide. Fuel 90:1963–1967
95. Uprety BK, Chaiwong W, Ewelike C, Rakshit SK (2016) Biodiesel production using heterogeneous catalysts including wood ash and the importance of enhancing byproduct glycerol purity. Energy Convers Manage 115:191–199
96. Li H, Niu S, Lu C, Li J (2016) Calcium oxide functionalized with strontium as heterogeneous transesterification catalyst for biodiesel production. Fuel 176:63–71
97. Chen G, Shan R, Shi J, Yan B (2014) Ultrasonic-assisted production of biodiesel from transesterification of palm oil over ostrich eggshell-derived CaO catalysts. Bioresource Technol 171:428–432
98. Sun H, Hu K, Lou H, Zheng X (2008) Biodiesel production from transesterification of rapeseed oil using KF/Eu_2O_3 as a catalyst. Energy Fuels 22:2756–2760
99. Malins K, Brinks J, Kampars V, Malina I (2016) Esterification of rapeseed oil fatty acids using a carbon-based heterogeneous acid catalyst derived from cellulose. Appl Catal, A 519:99–106
100. Rashid U, Anwar F, Knothe G (2009) Evaluation of biodiesel obtained from cottonseed oil. Fuel Process Technol 90:1157–1163
101. Chen H, Peng B, Wang D, Wang J (2007) Biodiesel production by the transesterification of cottonseed oil by solid acid catalysts. Front Chem Eng China 1:11–15
102. Lingfeng C, Guomin X, Bo X, Guangyuan T (2007) Transesterification of cottonseed oil to biodiesel by using heterogeneous solid basic catalysts. Energy Fuels 21:3740–3743
103. Rashid U, Anwar F, Ansari TM, Arif M, Ahmad M (2009) Optimization of alkaline transesterification of rice bran oil for biodiesel production using response surface methodology. J Chem Technol Biotechnol 84:1364–1370
104. Baskar G, Aberna Ebenezer Selvakumari I, Aiswarya R (2018) Biodiesel production from castor oil using heterogeneous Ni doped ZnO nanocatalyst. Bioresource Technol 250:793–798
105. Perin G, Álvaro G, Westphal E, Viana LH, Jacob RG, Lenardão EJ, D'Oca MGM (2008) Transesterification of castor oil assisted by microwave irradiation. Fuel 87:2838–2841

106. Yuan H, Yang BL, Zhu GL (2009) Synthesis of biodiesel using microwave absorption catalysts. Energy Fuels 23:548–552
107. Soriano NU, Venditti R, Argyropoulos DS (2009) Biodiesel synthesis via homogeneous Lewis acid-catalyzed transesterification. Fuel 88:560–565
108. Korkut I, Bayramoglu M (2018) Selection of catalyst and reaction conditions for ultrasound assisted biodiesel production from canola oil. Renew Energy 116:543–551
109. Alsharifi M, Znad H, Hena S, Ang M (2017) Biodiesel production from canola oil using novel Li/TiO_2 as a heterogeneous catalyst prepared via impregnation method. Renew Energy 114:1077–1089
110. Zhu H, Wu Z, Chen Y, Zhang P, Duan S, Liu X, Mao Z (2006) Preparation of biodiesel catalyzed by solid super base of calcium oxide and its refining process. Chin J Catal 27:391–396
111. Taufiq-Yap YH, Lee HV, Hussein MZ, Yunus R (2011) Calcium-based mixed oxide catalysts for methanolysis of Jatropha curcas oil to biodiesel. Biomass Bioenergy 35:827–834
112. Vyas AP, Subrahmanyam N, Patel PA (2009) Production of biodiesel through transesterification of Jatropha oil using KNO_3/Al_2O_3 solid catalyst. Fuel 88:625–628
113. Sudsakorn K, Saiwuttikul S, Palitsakun S, Seubsai A, Limtrakul J (2017) Biodiesel production from Jatropha Curcas oil using strontium-doped CaO/MgO catalyst. J Environ Chem Eng 5:2845–2852
114. Choudhury HA, Goswami PP, Malani RS, Moholkar VS (2014) Ultrasonic biodiesel synthesis from crude Jatropha curcas oil with heterogeneous base catalyst: mechanistic insight and statistical optimization. Ultrason Sonochem 21:1050–1064
115. Badday AS, Abdullah AZ, Lee K-T (2014) Transesterification of crude Jatropha oil by activated carbon-supported heteropolyacid catalyst in an ultrasound-assisted reactor system. Renew Energy 62:10–17
116. Buasri A, Inkaew T, Kodephun L, Yenying W, Loryuenyong V (2015) Natural hydroxyapatite (NHAp) derived from pork bone as a renewable catalyst for biodiesel production via microwave irradiation. Key Eng Mater 659:216–220
117. Meher LC, Kulkarni MG, Dalai AK, Naik SN (2006) Transesterification of karanja (*Pongamia Pinnata*) oil by solid basic catalysts. Eur J Lipid Sci Technol 108:389–397
118. Madras G, Kolluru C, Kumar R (2004) Synthesis of biodiesel in supercritical fluids. Fuel 83:2029–2033
119. Dong T, Gao D, Miao C, Yu X, Degan C, Garcia-Pérez M, Rasco B, Sablani SS, Chen S (2015) Two-step microalgal biodiesel production using acidic catalyst generated from pyrolysis-derived bio-char. Energy Convers Manage 105:1389–1396
120. Tran TTV, Kaiprommarat S, Kongparakul S, Reubroycharoen P, Guan G, Nguyen MH, Samart C (2016) Green biodiesel production from waste cooking oil using an environmentally benign acid catalyst. Waste Manage 52:367–374
121. Jacobson K, Gopinath R, Meher LC, Dalai AK (2008) Solid acid catalyzed biodiesel production from waste cooking oil. Appl Catal, B 85:86–91
122. González ME, Cea M, Reyes D, Romero-Hermoso L, Hidalgo P, Meier S, Benito N, Navia R (2017) Functionalization of biochar derived from lignocellulosic biomass using microwave technology for catalytic application in biodiesel production. Energy Convers Manage 137:165–173
123. Schreiber MW, Rodriguez-Niño D, Gutiérrez OY, Lercher JA (2016) Hydrodeoxygenation of fatty acid esters catalyzed by Ni on nano-sized MFI type zeolites. Catal Sci Technol 6:7976–7984
124. Cao X, Long F, Zhai Q, Liu P, Xu J, Jiang J (2020) Enhancement of fatty acids hydrodeoxygenation selectivity to diesel-range alkanes over the supported Ni-MoO_x catalyst and elucidation of the active phase. Renew Energy 162:2113–2125
125. Kon K, Onodera W, Takakusagi S, Shimizu K (2014) Hydrodeoxygenation of fatty acids and triglycerides by Pt-loaded Nb_2O_5 catalysts. Catal Sci Technol 4:3705–3712
126. Jiraroj D, Jirarattanapochai O, Anutrasakda W, Samec JSM, Tungasmita DN (2021) Selective decarboxylation of biobased fatty acids using a Ni-FSM-16 catalyst. Appl Catal B 291:120050

127. Wu J, Shi J, Fu J, Leidl JA, Hou Z, Lu X (2016) Catalytic decarboxylation of fatty acids to aviation fuels over nickel supported on activated carbon. Sci Rep 6:27820
128. Mäki-Arvela P, Kubickova I, Snåre M, Eränen K, Murzin DY (2007) Catalytic deoxygenation of fatty acids and their derivatives. Energy Fuels 21:30–41
129. He S, Zuur K, Santosa DS, Heeres A, Liu C, Pidko E, Heeres HJ (2021) Catalytic conversion of pure glycerol over an un-modified H-ZSM-5 zeolite to bio-based aromatics. Appl Catal B 281:119467
130. Zacharopoulou V, Vasiliadou ES, Lemonidou AA (2015) One-step propylene formation from bio-glycerol over molybdena-based catalysts. Green Chem 17:903–912
131. Cornejo A, Barrio I, Campoy M, Lázaro J, Navarrete B (2017) Oxygenated fuel additives from glycerol valorization. Main production pathways and effects on fuel properties and engine performance: a critical review. Renew Sustain Energy Rev 79:1400–1413
132. Deneyer A, Ennaert T, Cavents G, Dijkmans J, Vanneste J, Courtin CM, Dusselier M, Sels BF (2016) Compositional and structural feedstock requirements of a liquid phase cellulose-to-naphtha process in a carbon- and hydrogen-neutral biorefinery context. Green Chem 18:5594–5606.
133. De S (2022) Production of long-chain hydrocarbon biofuels from biomass-derived platform chemicals: catalytic approaches and challenges. In: Maity SK, Gayen K, Bhowmick TK (eds) Hydrocarbon biorefinery. Elsevier, pp 327–354
134. Rosatella AA, Simeonov SP, Frade RFM, Afonso CAM (2011) 5-Hydroxymethylfurfural (HMF) as a building block platform: biological properties, synthesis and synthetic applications. Green Chem 13:754–793
135. Delbecq F, Wang Y, Muralidhara A, El Ouardi K, Marlair G, Len C (2018) Hydrolysis of hemicellulose and derivatives—a review of recent advances in the production of furfural. Front Chem 6:146
136. Dutta S, Bhat NS (2021) Recent advances in the value addition of biomass-derived levulinic acid: a review focusing on its chemical reactivity patterns. ChemCatChem 13:3202–3222
137. Yan L, Yao Q, Fu Y (2017) Conversion of levulinic acid and alkyl levulinates into biofuels and high-value chemicals. Green Chem 19:5527–5547
138. Raspolli Galletti AM, Antonetti C, Fulignati S, Licursi D (2020) Direct alcoholysis of carbohydrate precursors and real cellulosic biomasses to alkyl levulinates: a critical review. Catalysts 10:1221
139. Ahmad E, Imteyaz Alam M, Pant KK, Ali Haider M (2016) Catalytic and mechanistic insights into the production of ethyl levulinate from biorenewable feedstocks. Green Chem 18:4804–4823
140. Mäki-Arvela P, Ruiz D, Murzin DY (2021) Catalytic hydrogenation/hydrogenolysis of 5-hydroxymethylfurfural to 2,5-dimethylfuran. ChemSusChem 14:150–168.
141. Vinod N, Dutta S (2021) Energy densification of biomass-derived furfurals to furanic biofuels by catalytic hydrogenation and hydrodeoxygenation reactions. Sustain Chem 2:521–549
142. Qian Y, Zhu L, Wang Y, Lu X (2015) Recent progress in the development of biofuel 2,5-dimethylfuran. Renew Sustain Energy Rev 41:633–646
143. Chen B, Li F, Huang Z, Yuan G (2017) Carbon-coated Cu-Co bimetallic nanoparticles as selective and recyclable catalysts for production of biofuel 2,5-dimethylfuran. Appl Catal, B 200:192–199
144. Solanki BS, Rode CV (2019) Selective hydrogenolysis of 5-(hydroxymethyl)furfural over Pd/C catalyst to 2,5-dimethylfuran. J Saudi Chem Soc 23:439–451
145. Sarkar C, Koley P, Shown I, Lee J, Liao Y-F, An K, Tardio J, Nakka L, Chen K-H, Mondal J (2019) Integration of interfacial and alloy effects to modulate catalytic performance of metal–organic-framework-derived Cu–Pd nanocrystals toward hydrogenolysis of 5-hydroxymethylfurfural. ACS Sustainable Chem Eng 7:10349–10362
146. Goyal R, Sarkar B, Bag A, Siddiqui N, Dumbre D, Lucas N, Bhargava SK, Bordoloi A (2016) Studies of synergy between metal–support interfaces and selective hydrogenation of HMF to DMF in water. J Catal 340:248–260

147. Nishimura S, Ikeda N, Ebitani K (2014) Selective hydrogenation of biomass-derived 5-hydroxymethylfurfural (HMF) to 2,5-dimethylfuran (DMF) under atmospheric hydrogen pressure over carbon supported PdAu bimetallic catalyst. Catal Today 232:89–98
148. Chen M-Y, Chen C-B, Zada B, Fu Y (2016) Perovskite type oxide-supported Ni catalysts for the production of 2,5-dimethylfuran from biomass-derived 5-hydroxymethylfurfural. Green Chem 18:3858–3866
149. Feng L, Li X, Lin Y, Liang Y, Chen Y, Zhou W (2020) Catalytic hydrogenation of 5-hydroxymethylfurfural to 2,5-dimethylfuran over Ru based catalyst: effects of process parameters on conversion and products selectivity. Renew Energy 160:261–268
150. Gao Z, Li C, Fan G, Yang L, Li F (2018) Nitrogen-doped carbon-decorated copper catalyst for highly efficient transfer hydrogenolysis of 5-hydroxymethylfurfural to convertibly produce 2,5-dimethylfuran or 2,5-dimethyltetrahydrofuran. Appl Catal B 226:523–533
151. Li J, Song Z, Hou Y, Li Z, Xu C, Liu C-L, Dong W-S (2019) Direct production of 2,5-dimethylfuran with high yield from fructose over a carbon-based solid acid-coated CuCo bimetallic catalyst. ACS Appl Mater Interfaces 11:12481–12491
152. Dutta S, Mascal M (2014) Novel pathways to 2,5-dimethylfuran via biomass-derived 5-(chloromethyl)furfural. ChemSuschem 7:3028–3030
153. Chen S, Ciotonea C, De Oliveira VK, Jérôme F, Wojcieszak R, Dumeignil F, Marceau E, Royer S (2020) Hydroconversion of 5-hydroxymethylfurfural to 2,5-dimethylfuran and 2,5-dimethyltetrahydrofuran over non-promoted Ni/SBA-15. ChemCatChem 12:2050–2059
154. Grochowski MR, Yang W, Sen A (2012) Mechanistic study of a one-step catalytic conversion of fructose to 2,5-dimethyltetrahydrofuran. Chem Eur J 18:12363–12371
155. Jackson MA, Appell M, Blackburn JA (2015) Hydrodeoxygenation of fructose to 2,5-dimethyltetrahydrofuran using a sulfur poisoned Pt/C catalyst. Ind Eng Chem Res 54:7059–7066
156. Liu X, Wang R (2018) Upgrading of carbohydrates to the biofuel candidate 5-ethoxymethylfurfural (EMF). Int J Chem Eng 2018:e2316939
157. Natsir TA, Shimazu S (2020) Fuels and fuel additives from furfural derivatives via etherification and formation of methylfurans. Fuel Process Technol 200:106308
158. Dong F, Zhu Y, Ding G, Cui J, Li X, Li Y (2015) One-step conversion of furfural into 2-methyltetrahydrofuran under mild conditions. ChemSuschem 8:1534–1537
159. Upare PP, Lee J-M, Hwang YK, Hwang DW, Lee J-H, Halligudi SB, Hwang J-S, Chang J-S (2011) Direct hydrocyclization of biomass-derived levulinic acid to 2-methyltetrahydrofuran over nanocomposite copper/silica catalysts. ChemSuschem 4:1749–1752
160. Yan K, Jarvis C, Gu J, Yan Y (2015) Production and catalytic transformation of levulinic acid: a platform for speciality chemicals and fuels. Renew Sustain Energy Rev 51:986–997
161. Martin Alonso D, Wettstein SG, Dumesic JA (2013) Gamma-valerolactone, a sustainable platform molecule derived from lignocellulosic biomass. Green Chem 15:584–595
162. Yan K, Yang Y, Chai J, Lu Y (2015) Catalytic reactions of gamma-valerolactone: a platform to fuels and value-added chemicals. Appl Catal B 179:292–304
163. James OO, Maity S (2022) Carbon-carbon (CC) bond forming reactions for the production of hydrocarbon biofuels from biomass-derived compounds. In: Maity SK, Gayen K, Bhowmick TK (eds) Hydrocarbon biorefinery. Elsevier, pp 297–325
164. Su M, Li W, Zhang T, Xin H, Li S, Fan W, Ma L (2017) Production of liquid fuel intermediates from furfural via aldol condensation over Lewis acid zeolite catalysts. Catal Sci Technol 7:3555–3561
165. Li G, Li N, Wang X, Sheng X, Li S, Wang A, Cong Y, Wang X, Zhang T (2014) Synthesis of diesel or jet fuel range cycloalkanes with 2-methylfuran and cyclopentanone from lignocellulose. Energy Fuels 28:5112–5118
166. Li G, Li N, Yang J, Wang A, Wang X, Cong Y, Zhang T (2013) Synthesis of renewable diesel with the 2-methylfuran, butanal and acetone derived from lignocellulose. Bioresour Technol 134:66–72
167. Mascal M, Dutta S, Gandarias I (2014) Hydrodeoxygenation of the angelica lactone dimer, a cellulose-based feedstock: simple, high-yield synthesis of branched C_7–C_{10} gasoline-like hydrocarbons. Angew Chem Int Ed 53:1854–1857

168. Chang F, Dutta S, Mascal M (2017) Hydrogen-economic synthesis of gasoline-like hydrocarbons by catalytic hydrodecarboxylation of the biomass-derived angelica lactone dimer. ChemCatChem 9:2622–2626
169. Li Z, Otsuki AL, Mascal M (2018) Production of cellulosic gasoline via levulinic ester self-condensation. Green Chem 20:3804–3808
170. Wu L, Mascal M, Farmer TJ, Arnaud SP, Wong Chang M-A (2017) Electrochemical coupling of biomass-derived acids: new C8 platforms for renewable polymers and fuels. ChemSuschem 10:166–170
171. Haq I, Mazumder P, Kalamdhad AS (2020) Recent advances in removal of lignin from paper industry wastewater and its industrial applications—a review. Bioresour Technol 312:123636
172. Katahira R, Elder TJ, Beckham GT (2018) A brief introduction to lignin structure. 1–20.
173. Lora JH, Glasser WG (2002) Recent industrial applications of lignin: A sustainable alternative to nonrenewable materials. J Polym Environ 10:39–48
174. Bajwa DS, Pourhashem G, Ullah AH, Bajwa SG (2019) A concise review of current lignin production, applications, products and their environmental impact. Ind Crops Prod 139:111526
175. Schutyser W, Renders T, Van de Bosch S, Koelewijn S-F, Beckham GT, Sels BF (2018) Chemicals from lignin: an interplay of lignocellulose fractionation, depolymerisation, and upgrading. Chem Soc Rev 47:852–908
176. Sun Z, Fridrich B, de Santi A, Elangovan S, Barta K (2018) Bright side of lignin depolymerization: Toward new platform chemicals. Chem Rev 118:614–678
177. Jiang L, Guo H, Li C, Zhou P, Zhang Z (2019) Selective cleavage of lignin and lignin model compounds without external hydrogen, catalyzed by heterogeneous nickel catalysts. Chem Sci 10:4458–4468
178. Konnerth H, Zhang J, Ma D, Prechtl MHG, Yan N (2015) Base promoted hydrogenolysis of lignin model compounds and organosolv lignin over metal catalysts in water. Chem Eng Sci 123:155–163
179. Venkatesan K, Krishna JVJ, Anjana S, Selvam P, Vinu R (2021) Hydrodeoxygenation kinetics of syringol, guaiacol and phenol over H-ZSM-5. Catal Commun 148:106164
180. Luo Z, Lu K, Yang Y, Li S, Li G (2019) Catalytic fast pyrolysis of lignin to produce aromatic hydrocarbons: optimal conditions and reaction mechanism. RSC Adv 9:31960–31968
181. Cao Z, Dierks M, Clough MT, Daltro de Castro IB, Rinaldi R (2018) A convergent approach for a deep converting lignin-first biorefinery rendering high-energy-density drop-in fuels. Joule 2:1118–1133
182. Hu X, Gholizadeh M (2019) Biomass pyrolysis: a review of the process development and challenges from initial researches up to the commercialisation stage. J Energy Chem 39:109–143
183. Wang G, Dai Y, Yang H, Xiong Q, Wang K, Zhou J, Li Y, Wang S (2020) A review of recent advances in biomass pyrolysis. Energy Fuels 34:15557–15578
184. Guedes RE, Luna AS, Torres AR (2018) Operating parameters for bio-oil production in biomass pyrolysis: a review. J Anal Appl Pyrolysis 129:134–149
185. Lv P, Yuan Z, Ma L, Wu C, Chen Y, Zhu J (2007) Hydrogen-rich gas production from biomass air and oxygen/steam gasification in a downdraft gasifier. Renew Energy 32:2173–2185
186. Arnold RA, Hill JM (2019) Catalysts for gasification: a review. Sustain Energy Fuels 3:656–672
187. Unruh D, Pabst K, Schaub G (2010) Fischer-Tropsch synfuels from biomass: maximizing carbon efficiency and hydrocarbon yield. Energy Fuels 24:2634–2641
188. Mandal S, Daggupati S, Majhi S, Thakur S, Bandyopadhyay R, Das AK (2019) Catalytic gasification of biomass in dual-bed gasifier for producing tar-free syngas. Energy Fuels 33:2453–2466
189. de Lasa H, Salaices E, Mazumder J, Lucky R (2011) Catalytic steam gasification of biomass: catalysts, thermodynamics and kinetics. Chem Rev 111:5404–5433

Experimental Investigation of Used Vegetable Oil-Diesel Blends as Alternative to Fossil Fuel in Compression Ignition Engine

Joseph O. Dirisu, Sunday O. Oyedepo, Precious I. Airhihen, Damola S. Adelekan, Uyi K. Efemwenkiekie, and Anish Khan

Abstract Partial replacement of crude oil is required to solve the challenge of depletion of fossil fuel and climate change. Environmental problems occur from the increasing emissions of harmful pollutants and greenhouse gases from the combustion of fossil fuels. Hence, the use of clean energy sources including biodiesel is crucial in this century. In this study, the use of waste vegetable oil blends with diesel (D95B5, D90B10, D85B15, and D80B20) as a retrofit for pure Diesel (D100) was investigated experimentally in a single-cylinder two-stroke diesel engine. Engine performance indices such as engine torque, engine power, brake mean effective pressure (BMEP), exhaust gas temperature, volumetric efficiency, air mass flow, fuel mass flow, air/fuel flow ratio, and thermal efficiency were assessed for 8, 16, and 24 ml fuel blends. The volume of fuel blends in the test rig was investigated under constant temperature (34 °C) conditions. Experimental results revealed that the blended fuels worked effectively in the rig and D85B15 gave the best performance within the tested diesel engine. This study established the possibility of partial replacement of pure diesel with fuel blends of diesel and used vegetable oil in internal combustion engine.

Keywords Air-fuel ratio · Used vegetable oil · Fuel · Diesel engine · Engine performance parameters

J. O. Dirisu · S. O. Oyedepo (✉) · P. I. Airhihen · D. S. Adelekan · U. K. Efemwenkiekie
Department of Mechanical Engineering, Covenant University, Ota P.M.B 1023, Ogun State, Nigeria
e-mail: Sunday.oyedepo@covenantuniversity.edu.ng

A. Khan
Chemistry Department, Center of Excellence for Advanced Materials Research (CEAMR), Faculty of Science, King Abdulaziz University, Jeddah, Saudi Arabia

A. Khan et al. (eds.), *Waste to Biofuel Technology*, Sustainable Materials and Technology, https://doi.org/10.1007/978-981-97-4561-6_4

1 Introduction

Waste is an unexploited material or a residue from a product in which its application is not yet identified. The importance of reusing waste cannot be underestimated. Waste contributes in conserving energy associated with the production of raw material as observed in the construction industries. Its use reduces the burden of occupying the land mass [1–3]. Recycling waste such as used oil is preferable to its disposal as it provides environmental benefits such as pollution reduction and balance in the ecosystem [4]. Used oil can also be used as an additive to diesel base oil to generate energy in the compression ignition engine [5, 6]. The subject matter of energy has been of great interest to the international community, several economic summits, and international conferences. It is a vital indicator of where the economy of a nation is heading. Energy can be likened to be the road map that shows the growth and progress of a nation's economy. In all parts of the world today, the demand for energy is increasing almost daily. Therefore, sustainable development, greenhouse gas emissions, and energy supply/demand have developed into 'fervent' issues in the past decades at international and regional levels involving various actors from policymakers to scientists. These actors have addressed their concerns through studies and reports [7].

1.1 Global Energy Consumption and CO_2 Emissions

Energy is a focal contributor to the economy and development of a nation. Nearly all sectors of government employ energy to operate such as manufacturing industries, transportation, residential buildings and offices, power generation and distribution, health, and education amongst others. There are various sources of energy which are specifically applied by countries in the world such as geothermal, nuclear fuels, biomass, fossil fuels, natural gas, coal, etc. In 2018, fossil fuel stands as the most widely used energy source with well above 82% of global energy consumption [8].

Global energy consumption has risen rapidly over the past century, due to population and increasing prosperity. The increase demand cuts across all energy sources—coal, oil, natural gas, nuclear energy, and renewables. Such growth in the global energy system reflects broadly positive trends of higher living standards, decreasing global poverty, and longer livelihoods. However, rising fossil fuel consumption is responsible for global climate change and major environmental challenges. To address these challenges, the global energy system will need to undergo a clean energy transition, whereby sources of energy that emit greenhouse gases are replaced by increasingly cleaner sources [9].

In view of the fact that energy demand is expected to grow by about 27%, or 3,743 million tons of oil equivalent (mtoe), worldwide from 2017 to 2040. According to IEA's New Policies Scenarios, carbon dioxide emissions from energy consumption would grow from 32.6 to 35.9 gigatons of carbon dioxide between 2017 and 2040,

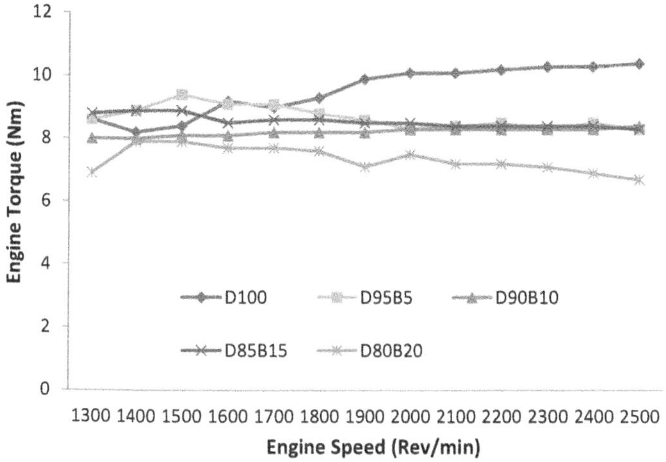

Fig. 1 Engine torque and speed variation within the test rig

a rise of 10%. The effect of this is on the ecosystem under threat from exploration, production, and the utilization of fossil fuel causing continued global warming and exhaustion of the ozone layer amplified by the discharged greenhouse gases. Therefore, there is a need for a new energy policy that will encourage the use of biodiesel blends in diesel engines for transport or power sector [10].

The demand and consumption of fossil fuel is increasing in developing countries such as Nigeria. Nigeria is acclaimed the prominent economy and also the prosperous oil reserve in Africa. The country has the prevalent gas customer and manufacturer of West Africa. The population of the country is about 200 million with a Gross National Product (GDP) of 460 billion USD [11, 12]. Population surge tend to increase on energy consumption in the world [13]. Nigeria currently depends on fossil fuel for power generation and in transportation sector. Figure 1 presents the rising demand for fossil fuel energy, an indicator of the importance of energy consumption by the populace.

1.2 Fuel Alternative Partial Replacement of Pure Diesel, D100

Fossil fuel in the form of crude oil meets Nigerian principal energy consumption. There has been concern for over depending on it and the negative environmental impact suffered from its extraction. Hence there is a need for alternative fuel or partially blending the fossil fuel with useful waste [14]. The intention of an alternative fuel drive is based on decarbonization of the energy sector, future depletion of fossil fuel, improved efficiency of the ICE system and reduction of noxious substances to the ecosystem [15]. Various types of alternative fuel exist that are successfully employed to power engines such as biomethanol-biodiesel diesel blend utilized in marine engineering [16]. Waste lubricating oil was applied to marine diesel engine as an alternative to fossil fuel [17]. Ethanol, natural gas, diesel, and gasoline were combined to fuel cars [18]. Cooking oil residue and corn biodiesel were used to run diesel engine [19]. Ethanol-RME-Diesel blend was employed on a diesel engine operating with a single cylinder [20]. Compression ignition engine (CIE) is observed as an adaptive ICE system for testing alternative fuel. Improvised blends tested on CIE are not limited to mesua ferrea linn oil, n-butanol vegetable oil blends, alcohol vegetable oil biodiesel-blend, linseed oil-diesel blend, jatrophacurcas vegetable oil, cotton seed oil biodiesel, castor oil/pure diesel blend, algae oil, fennel vegetable oil, nanomodifier/hydrotreated vegetable oil, jojoba oil blend with butanol additive and palm oil-diesel blend [21–34].

This research aims at studying the engine performance of a diesel engine by partial replacement of diesel with vegetable oil residue without any modification to the diesel engine.

2 Materials and Methods

Versatile Data Acquisition System (VDAS), TD200 Small Engine testbed set fixed with TQ P860 Dynamometer control pack, 25 L waste Vegetable Oil, 25 L of Diesel, fine sieves of sizes 300, 200, 100, and 50 μm, measuring cylinders of 500 ml capacity and thermocouple K were employed for investigating the effectiveness of diesel motor by engaging vegetable oil-diesel blends. Dynamometer was used for determining the engine torque, power, and Speed and the loading of the engine (i.e., increasing and decreasing the load). The dynamometer displays the values of the engine speed, torque, power, as well as the ambient and exhaust temperatures, air bore differential pressure, and ambient temperature. A bulb with a capacity to display the flow of the fuel through an 8 and 16 ml compartment; hence giving the value for the time for 24 ml (i.e., 8 ml + 16 ml) was also attached to the dynamometer.

The testbed works with a VDAS. It also serves as a base upon which the engine and the dynamometer was mounted. The two-stroke engine was used to analyze the behaviour of the blends. These blends were compared with a control sample (i.e., a sample which is exclusively 100% diesel). A control sample of 3000 ml was accurately measured using a measuring cylinder and poured into a tank of 3000 ml capacity. The 3000 ml (3 L) capacity tank was mounted on a structure that is connected via pipes to a digital display dynamometer. The entire structure was an assembly made of a Versatile Data Acquisition System, a TD200 Small Engine testbed set mounted with TQ P860 Dynamometer control pack, and accessories made of pipes connected to a water source to convey water from a tap to the testbed.

2.1 Experimental Setup

The experiment was conducted in an ambient temperature range of 25–32 °C measured using a digital Thermocouple K. Experimental test rig employed for this investigation was a TD 200 two-stroke diesel engine observed under different Speed (1300, 1400...2500 rpm). The rig was equipped with a testbed having a dynamometer and VDAS for engine speed variation and parameters (Speed, torque, power, exhaust gas temperature, differential box pressure, and ambient temperature) observation.

In this experiment, two types of data were generated, namely; Primary data and Secondary data. Primary data were obtained directly using the VDAS. This data includes engine speed, torque, power, exhaust gas temperature, air bore differential pressure, fuel drainage time, ambient pressure, and ambient air temperature at the exhaust inlet. The engine speed is simply the time rate of engine shaft rotation and was directly computed by VDAS. Furthermore, the engine speed was the dependent variable for obtaining all other above-listed primary data collected from VDAS that works primarily on thermodynamic equation functions and instrumentation of the test rig temperature and pressure gauges. The secondary data were derived by the application of under-listed performance analyses. Data including air mass flow, fuel mass flow, mass air/fuel ratio, calculated volume, measured volume, and Heat of combustion of fuel were used to describe the fuel blends flow performance characteristics. In contrast, the blends heat transfers and mechanical characteristics performance analyses were induced from brake mean effective pressure and thermal Efficiency data, respectively.

2.2 Preparation of Used Vegetable-Diesel Oil Blends

Employed fuel blends consist of pure Diesel and used vegetable oil obtained locally from NNPC filling station and Covenant University cafeteria (Cafeteria 2), respectively. The used vegetable oil was sieved appropriately and mixed in proportions presented below:

i. 5% used vegetable oil and 95% diesel (D95B5)
ii. 10% used vegetable oil and 90% diesel (D90B10)
iii. 15% used vegetable oil and 85% diesel (D85B15)
iv. 20% used vegetable oil and 80% diesel (D80B20).

The under-listed volumetric calculations were used to develop various fuel blends:

i. $D95B5 = \frac{95}{100} \times 3000\,\text{ml} = 2850\ \text{ml}$
ii. $D90B10 = \frac{90}{100} \times 3000\,\text{ml} = 2700\ \text{ml}$
iii. $D85B15 = \frac{85}{100} \times 3000\,\text{ml} = 2550\ \text{ml}$
iv. $D90B10 = \frac{80}{100} \times 3000\ \text{ml} = 2400\ \text{ml}.$

Each sample was prepared by calculating the quantity of Diesel and used vegetable oil needed for each blend as indicated in Eqs. 1–10; pouring the measured amount of used vegetable oil through a piece of foam 10 mm thick placed on a sieve to filter the oil; mixing the Diesel and filtered used vegetable oil in the right proportions to get the exact blend. The blend was poured into the tank, and the engine was started. Adjustments were made to vary the quantity of water entering the machine to reduce or increase (regulate) the exhaust temperature, hence controlling the engine speed, power, and torque. The readings were then taken from the digital display mechanism of the dynamometer.

2.3 Engine Performance Evaluation

There were assumptions made to evaluate the performance of the fuel for ease of calculations such as the engine being in a steady-state condition. The reference state is defined as $T_0 = 293$ K and $P_0 = 1$ atm, the combustion air, and exhaust gases are ideal gas mixtures.

Air mass flow (m_a):

$$m_a = C_d \frac{\pi d^2}{4} \sqrt{\frac{2P_A \Delta P}{RT_A}} \tag{1}$$

where C is the calorific value of fuel in J/kg, Δp is a pressure change in Pa, d is the orifice diameter in m, R is the gas constant in J/kgK, P_A is ambient pressure Pa, and T_A is the ambient air temperature at the inlet in K.

Fuel mass flow (m_f):

The fuel mass flow of various fuel blends for tested engine speeds is represented as m_f, is calculated and measured with the equation,

$$m_f = \frac{V_f}{1000} \times \rho_f \tag{2}$$

where V_f is the fuel volume flow and ρ_f is the fuel density.

Mass air/fuel ratio which is an index of the air and the fuel flow within the test rig is represented as

$$\text{Mass air to fuel} = \frac{m_a}{m_f} \tag{3}$$

where m_a is the air mass flow and m_f is the fuel mass flow.

Specific fuel consumption (SFC):

This is defined as the fuel flow rate per unit output power.

$$\mathrm{SFC} = \frac{m_f \times 3600}{P/1000} \quad (4)$$

where m_f is the fuel mass flow and P is the mechanical power.

Measured volume: This is the experimental instantaneous fuel consumption of an engine. It is represented by M_v and can be calculated by the equation

$$Mv = \frac{m_{a \times R} \times T_A}{P_{A \times 100}} \quad (5)$$

Calculated volume (C_v) This is the direct function of the engine capacity. It is represented as C_v and calculated using the equation

$$C_v = \frac{E_c \times N}{\left(\frac{S}{2}\right) \times 60} \quad (6)$$

where E_c is the engine capacity, N is the engine speed and the number of strokes.

Volumetric Efficiency (V):

$$V = \frac{M_v}{C_v} \times 100 \quad (7)$$

Heat of Combustion of fuel $\left(H_f\right)$:

This is represented as (H_f), and it is a function of the fuel mass flow and the calorific value of the fuel. It is derived from the equation

$$\left(H_f\right) = m_f C_L \times 10^6 \quad (8)$$

where C_L is the calorific value given as 38.99 MJ·Kg

Thermal Efficiency (Δ_T):

$$\Lambda_T = \frac{P}{H_f} \times 100 \quad (9)$$

Brake Mean Effective Pressure (BMEP)

$$\mathrm{BMEP} = \frac{60 \times P \times \left(\frac{S}{2}\right)}{0.1 \times v \times E_c} \quad (10)$$

Air inlet Enthalpy (H_A)

$$\text{Air inlet Enthalpy}(H_A) = m_a C_p T_A \times 10^6 \quad (11)$$

3 Results and Discussion

3.1 Engine Torque Variation Characteristics

Torque is simply the product of applied force and radius about the axis of its rotation on a power-generating shaft. Figure 1 presents the torque generated at various operating Speed in the engine testbed working with varied volume ratios of vegetable oil and diesel blend. Obtained results show that torque increases with an increase in the engine for D100 and D90B10 blends while other blends, including D95B5, D85B15, and D80B20, showed an undulating increase and decrease of the torque with the increase in engine speed. The average torque obtained for pure and blends of fuel was 9.53, 8.69, 8.22, 8.55, and 7.36 Nm for D100, D95B5, D90B10, D85B15, and D80B20 fuel in the test rig. Percentage reduction in torque generation was obtained for all used vegetable oil and diesel fuel blends. The least percentage reduction was achieved using the B5 fuel blends, which had a 9.02% reduction while 10.33, 13.97, and 23.05% reductions were obtained using B15, B10, and B20 fuel blends respectively. This result is consistent with the results in the works of [35–37].

3.2 Results of Engine Power Variation Characteristics

Power illustrates the time rate of work done in a process. Figure 2 depicts the power and engine speed variation in the testbed for various fuel compositions within the testbed. The result analysis obtained showed an increase in power generated for increasing Speed within the testbed. Also, the highest average power was generated using pure diesel fuel, while the least was with the D80B20 fuel blends. Average power generated for pure and blends of fuel include; D95B5, D90B10, D85B15, and D80B20 were 1892.42, 1712.31, 1631.39, 1613.77, and 1453.62 W respectively. Fuel blends had 7.09, 11.48, 12.43, and 21.12% percentage reductions in power generated for D95B5, D90B10, D85B15, and D80B20, respectively, when compared to pure diesel fuel in the test rig. This is corresponding with the results in the works of [37, 38]. The reasons for this result are due to the decrease in air-fuel mixture temperature at the beginning of the combustion stroke resulting from the decrease in the lower heating value of the vegetable oil-diesel blends, as a consequent, combustion temperature decreases; the and lower viscosity of Diesel.

3.3 Brake Effective Mean Pressure Variation

The results of calculated variations in brake mean effective pressure using the employed fuel sources are shown in Fig. 3. The highest average brake means effective pressure (BMEP) of 2.50 MPa was obtained using D100 fuel while the least was

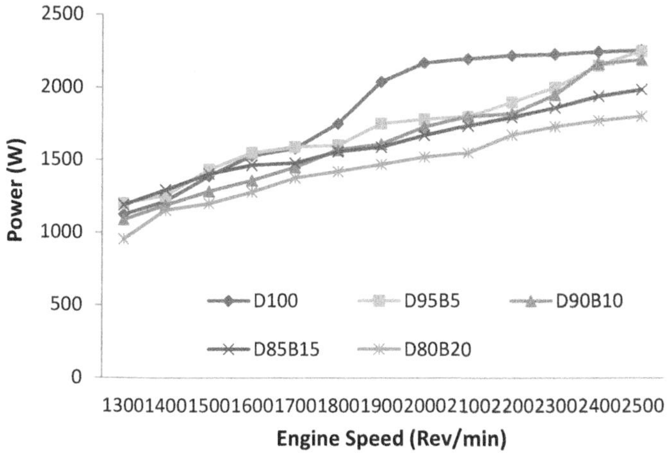

Fig. 2 Engine power variation with speed for pure and all fuel blends in the test rig

with the D80B20 fuel blends, which had 2.0 MPa average BMEP in the test rig. Also, all fuel blends had 6.20, 11.18, 11.04, and 20.29% percentage reduction in calculated brake mean effective pressures for D95B5, D90B10, D85B15, and D80B20 fuel blends, respectively. Furthermore, D95B5, D90B10, and D85B15 fuel blends had 2.34, 2.22, and 2.22 MPa, respectively. Calculated average BMEPs within the engine testbed.

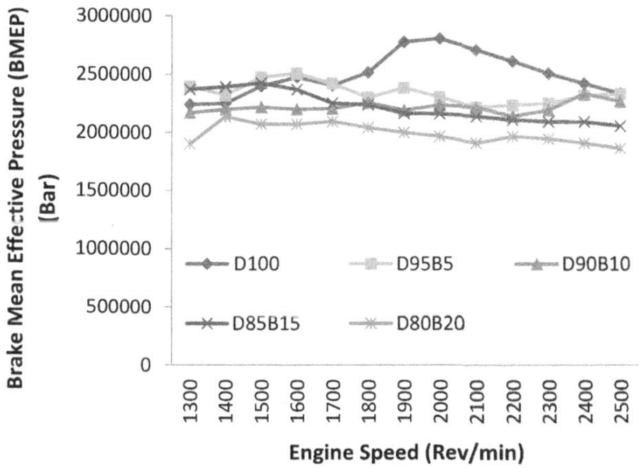

Fig. 3 Engine brake mean effective pressure variation with Speed for pure and all fuel blends in the test rig

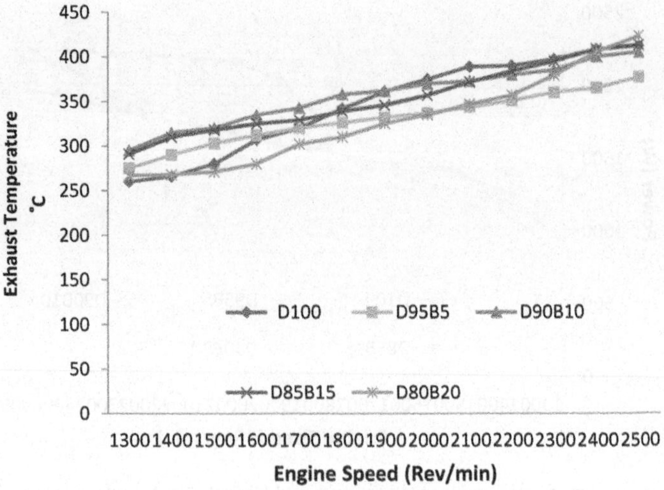

Fig. 4 Engine exhaust gas temperature with speed variation for pure and all fuel blends in the test rig

3.4 Results of Exhaust Gas Temperature

Exhaust gas is simply a by-product of combustion obtained from fuel and air reaction within the combustion chamber of an ICE engine. Figure 5 illustrates the mean temperature of the exhaust gases produced and channelled out through the exhaust manifold while using each type of fuel at various engine speeds. Works of the literature of [35] also reveal similar results (Fig. 4).

3.5 Results of Volumetric Efficiency

In Fig. 5, the average volumetric efficiency of all fuel blends type is shown. Results of calculation revealed that the maximum and lowest mean volumetric efficiency was obtained using the B15 fuel blends, which had 2.7% and D100 with 2.34% volumetric efficiencies, respectively. All other blends containing used vegetable oil had upper volumetric efficiencies than D100 with results including; 2.47% for D95B5, 2.52% for D90B10, and 2.63% for D85B15 fuel blends. The work of [35] had a similar result; the reason for this is since the induction air increases the volumetric efficiency as the load increases.

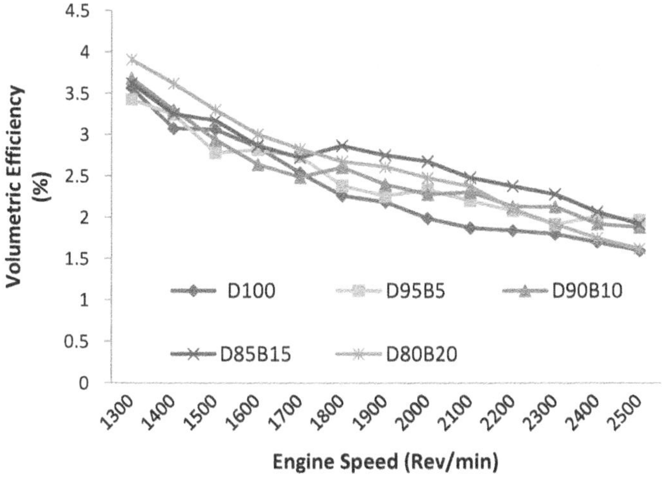

Fig. 5 Volumetric efficiency variation with engine speed for pure and all fuel blends in the test rig

3.6 Results of Air Mass Flow Variation

Calculated air mass flow using equation one at various engine speeds is as presented in Fig. 6. Reduction in average air mass flow values, when compared to pure D100, was obtained using D95B5 and D80B20 fuel blends in the rig, while increased average air mass flow values were achieved using D90B10 and D85B15 fuel blends. Attained average air mass flows were 300×10^{-6}, 286×10^{-6}, 301×10^{-6}, 313×10^{-6} and 299×10^{-6} kg/s for D100, D95B5, D90B10, D85B15, and D80B20 fuel types respectively. D95B5 and D80B20 fuel blends had 4.8 and 0.4%, lesser average fuel mass flow in the rig, and D90B10 and D85B15 fuel blends had 0.2 and 4.2%, more average fuel mass flow in the rig.

3.7 Results of 8 ml Volume Fuel Mass Flow Variation

Figure 7 shows the results obtained for fuel mass flow variations at different speeds for 8 ml volume quantity within the engine testbed using Eq. (2). All blends of fuel used had a reduction in average fuel mass flows when compared to pure D100 in the testbed. Results gotten for D100, D95B5, D90B10, D85B15 and D80B20 fuel types are; 164×10^{-6}, 144×10^{-6}, 148×10^{-6}, 144×10^{-6}, and 146×10^{-6} kg/s respectively. D95B5, D90B10, D85B15, and D80B20 fuel blends had 12.4, 9.72, 12.08, and 11.29% lesser average fuel mass flow in the rig.

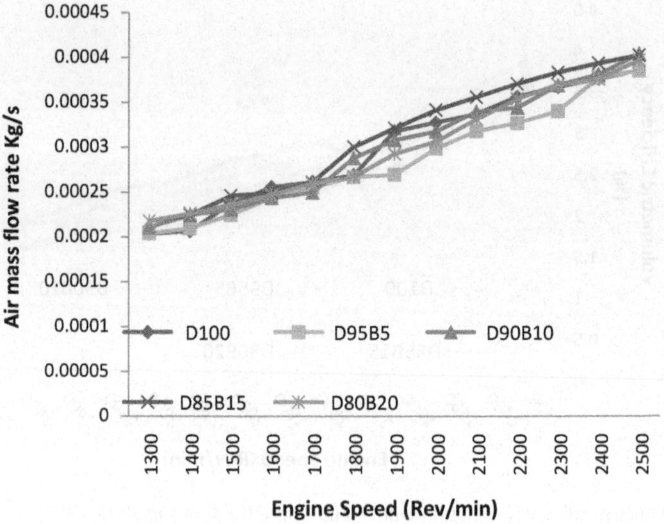

Fig. 6 Air mass flow with engine speed ratio for pure and all fuel blends in the test rig

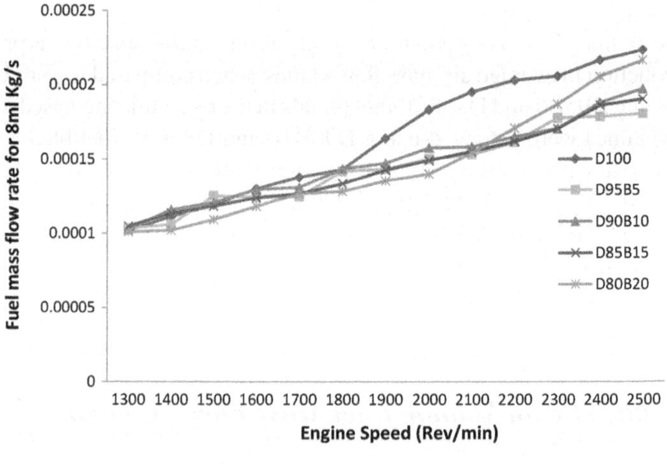

Fig. 7 Fuel mass flow for 8 ml fuel volume variation with engine speed ratio for pure and all fuel blends in the test rig

3.8 Results of 16 ml Volume Fuel Mass Flow Variation

Calculated fuel mass flow for 16 ml volume quantity at various engine speeds in the test rig is, as shown in Fig. 8. Similarly, all blends of fuel compositions had reduced average fuel flow for 16 ml fuel volume when compared to pure D100 fuel type in the testbed. Percentage reductions of 13.03, 9.29, 13.1, and 11.36% for D95B5,

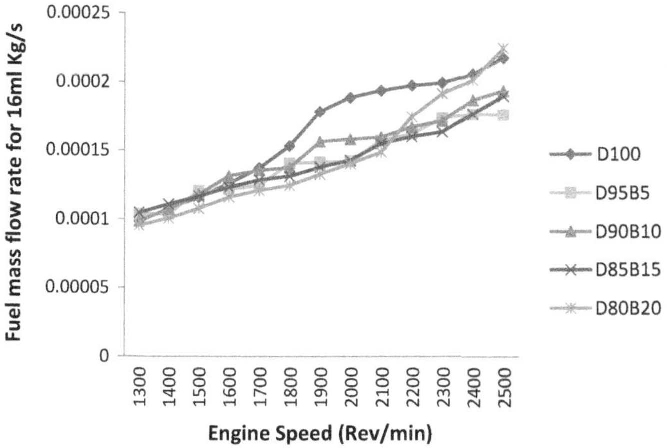

Fig. 8 Fuel mass flow for 16 ml fuel volume variation with engine speed ratio for pure and all fuel blends in the test rig

D90B10, D85B15, and D80B20 fuel compositions, respectively, were achieved in the rig. Also, their average fuel mass flow using 16 ml volume was 163×10^{-6}, 142×10^{-6}, 148×10^{-6}, 142×10^{-6} and 145×10^{-6} kg/s for D100, D95B5, D90B10, D85B15 and D80B20 fuel compositions respectively.

3.9 Results of 24 ml Volume Fuel Mass Flow Variation

In Fig. 9, the calculated volume fuel mass flow at the various engine is illustrated. All average and percentage reductions in fuel flow for 24 ml fuel volume in the test rig is the same to 16 ml fuel flow result when compared to pure D100 fuel type in the rig except for D85B15 fuel composition which had 143×10^{-6} kg/s and 12.56% average fuel mass flow and percentage reduction indexes.

3.9.1 Results of Air/Fuel Ratios for 8 ml Fuel Volume Variation

Calculated air/fuel ratios for 8 ml fuel volume at various engine speeds using Eq. (7) in the test rig results are presented in Fig. 10. All fuel blends had improved calculated air/fuel ratios of 7.18, 9.27, 16.69, and 12.22% for fuel blends, respectively. The average calculated air/fuel ratios obtained were 1.85, 1.98, 2.02, 2.16, and 2.08 for D100, D95B5, D90B10, D85B15, and D80B20 fuel types respectively.

Fig. 9 Fuel mass flow for 24 ml fuel volume variation with engine speed ratio for pure and all fuel blends in the test rig

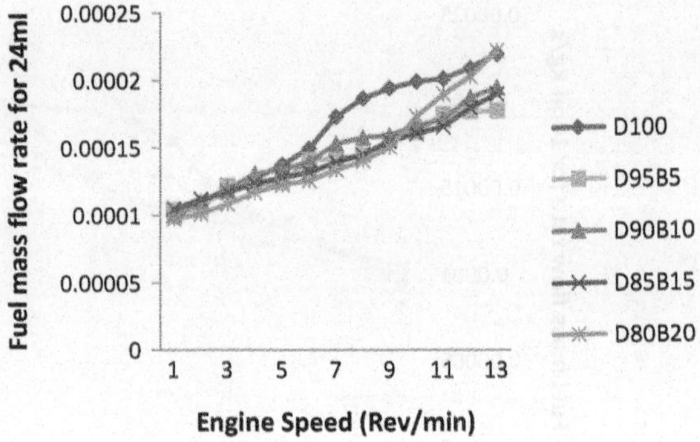

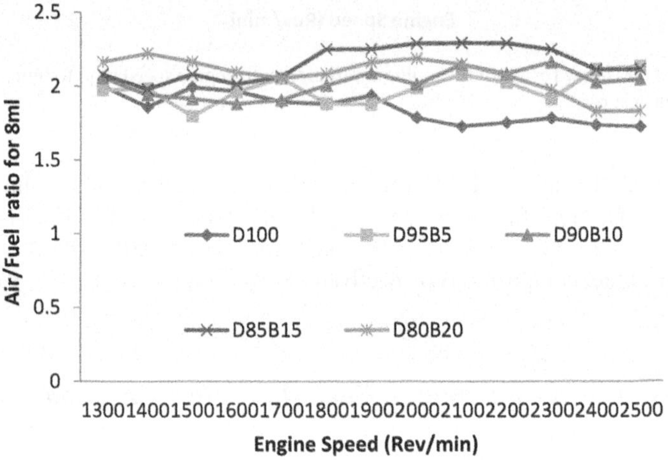

Fig. 10 Air/fuel ratio for 8 ml fuel volume variation with engine speed ratio for pure and all fuel blends in the test rig

3.9.2 Results of Air/Fuel Ratios for 16 ml Fuel Volume Variation

In Fig. 11, the result of the calculated air/fuel ratio for 16 ml fuel volume quantity at various engine speeds behaved similarly to 8 ml fuel volume quantity in the rig. Results revealed that the average calculated air/fuel ratios of 2.01, 2.03, 2.2, and 2.11 were obtained for D95B5, D90B10, D85B15, and D80B20 fuel blends respectively, which was higher than pure D100 which had 1.86 air/fuel ratio values. Furthermore, calculated percentage improvement using the fuel blends was 7.87, 8.96, 17.91, and 12.88%, respectively.

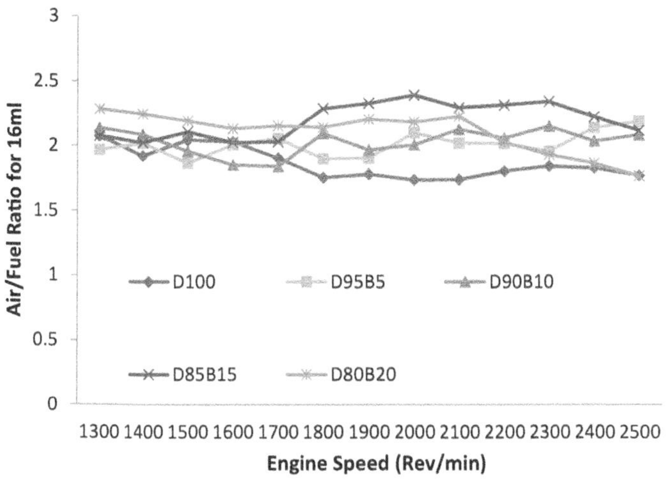

Fig. 11 Air/fuel ratio for 16 ml fuel volume variation with engine speed ratio for pure and all fuel blends in the test rig

3.9.3 Results of Air/Fuel Ratios for 24 ml Fuel Volume Variation

Results of calculated air/fuel ratios for 24 ml volume quantity at various engine speeds are as presented in Fig. 12. Obtained calculated average 24 ml fuel volume air/fuel ratios were 1.86, 2, 2.03, 2.19, and 2.1 for D100, D95B5, D90B10, D85B15, and D80B20 fuel blends. This shows all fuel blends similarly improved in their calculated average air/fuel ratios. Furthermore, 7.64 9.07, 17.5, and 12.66% percentage improvement was attained using D95B5, D90B10, D85B15, and D80B20 fuel blends in the rig.

3.9.4 Results of Thermal Efficiency Variation for 8 ml Volume

Figure 13 illustrates the thermal efficiency variation at different Speed for 8 ml volume for all tested fuel compositions. Results of average thermal Efficiency showed that only the B5 fuel blends recorded an improvement in thermal efficiency when compared to pure D100 fuel type. In contrast, other fuel blends had declined thermal efficiency, respectively. The increase in thermal Efficiency of B5 is due to the improvement of the combustion process on account of increased oxygen content in the fuels according to [39]. However, the drop in the thermal efficiency of other biodiesel fuel blends may be due to unfortunate spray characteristics which are due to higher viscosity and surface tension [40] and lower calorific value of biodiesels compared to that of Diesel [41]. The weak spray pattern affects the homogeneity of the air-fuel mixture, which in turn lower the heat release rate thereby reducing the thermal efficiency [39].

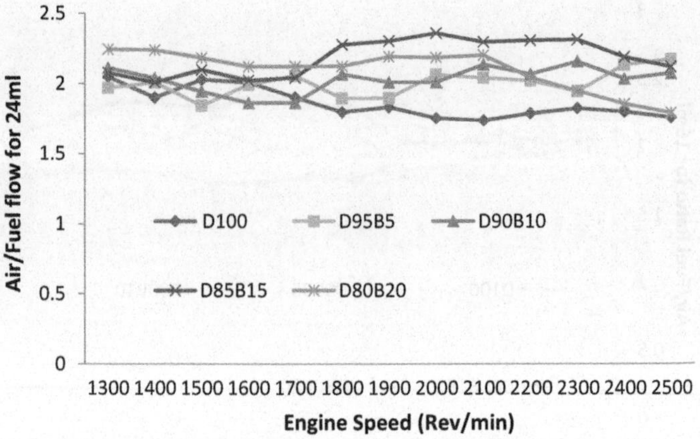

Fig. 12 Air/fuel ratio for 24 ml fuel volume variation with engine speed ratio for pure and all fuel blends in the test rig

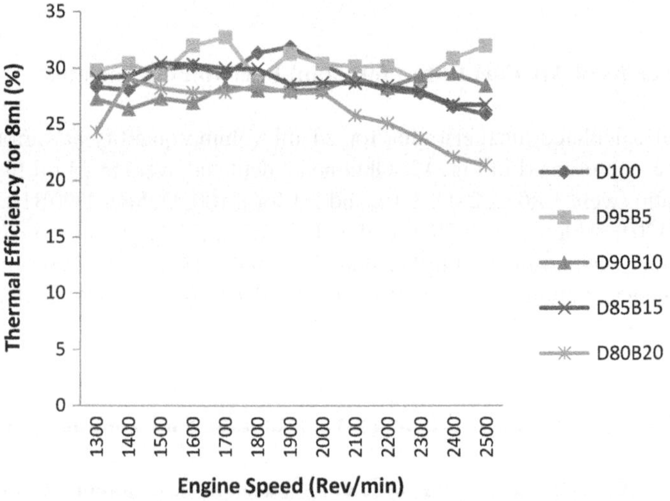

Fig. 13 Thermal Efficiency for 8 ml fuel volume variation with engine speed ratio for pure and all fuel blends in the test rig

3.9.5 Result of Thermal Efficiency Variation with Engine Speed for 16 ml Fuel Volume

Figure 14 presents the detailed illustration of thermal efficiency variations with engine speeds for 16 ml fuel volume quantity all varied composition of Diesel and used vegetable oil. Average thermal efficiency indexes obtained were 29.16, 30.97, 28.22, 29.35, and 26.47% for D100, D95B5, D90B10, D85B15, and D80B20 fuel

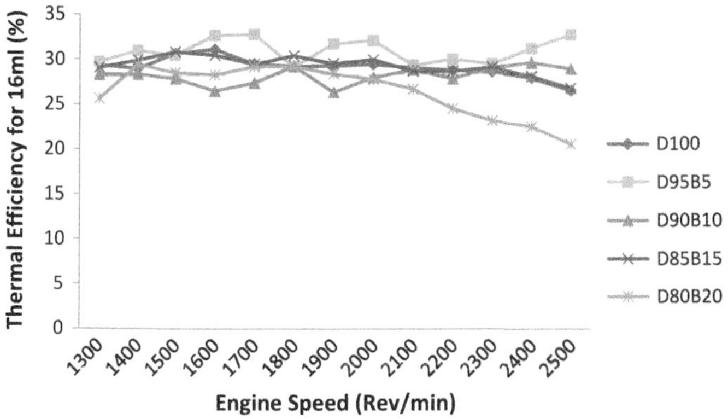

Fig. 14 Thermal Efficiency for 16 ml fuel volume variation with engine speed ratio for pure and all fuel blends in the test rig

blends respectively. Improvement of 6.24 and 0.66% in average thermal efficiency variation was obtained for D95B5 and D85B15 fuel blends respectively, while deterioration of 3.22 and 9.22% was achieved using D90B10 and D80B20 fuel blends respectively within the testbed.

3.9.6 Result of Thermal Efficiency Variation with Engine Speed for 24 ml Fuel Volume

Figure 15 similarly illustrates the thermal efficiency variation obtained at different engine speeds in the test rig using a 24 ml volume quantity of each fuel blends. The average thermal efficiency indexes gotten were 29.1, 30.83, 28.17, 29.18, and 26.35% for D100, D95B5, D90B10, D85B15, and D80B20 fuel blends correspondingly. Improvement of 5.95 and 0.29% in average thermal efficiency variation was obtained for D95B5 and D85B15 fuel blends while deterioration of 3.17 and 9.45% was achieved using D90B10 and D80B20 fuel blends respectively within the testbed.

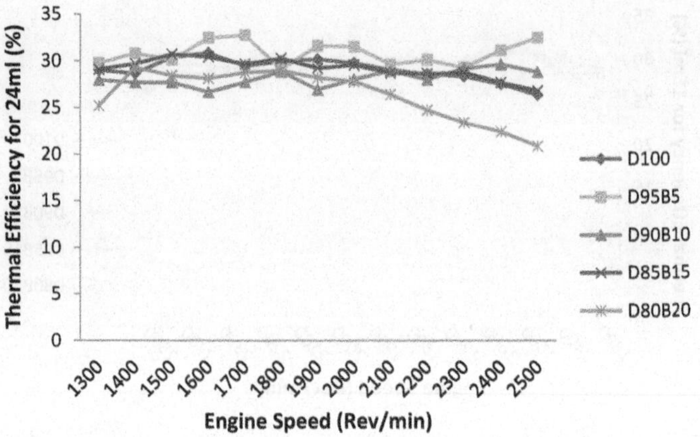

Fig. 15 Thermal efficiency for 24 ml fuel volume variation with engine speed ratio for pure and all fuel blends in the test rig

4 Conclusion

This study assessed the performance of used vegetable oil-diesel blends as an alternative to fossil fuel in compression ignition engine experimentally. Results of the study show that the used vegetable oil-diesel blends can serve in a replacement for pure diesel oil in internal combustion engine. As the fuel blends had 6.20, 11.18, 11.04, and 20.29% reduction in BMEP for D95B5, D90B10, D85B15, and D80B20, respectively in comparison to pure diesel. There was also a reduction in fuel mass flow rates with fuel blends in comparison to pure diesel. The air/fuel ratios of fuel blends increased by 7.18, 9.27, 16.69, and 12.22% for D95B5, D90B10, D85B15, and D80B20, respectively when compared to the pure diesel. Based on thermal efficiency, only D95B5 fuel blends had an improvement in thermal efficiency when compared to pure diesel (D100) fuel. Test results revealed that the blends worked well in the test rig, and D85B15 gave the optimal performance of the tested diesel engine. Conclusively, this experimental study shows that used vegetable oil-diesel blends can serve as a replacement to pure diesel (D100) in compression ignition engine.

Global energy demand and CO_2 emissions trends in the Stated Policies Scenario, 2019–2030.

In view of the fact that energy demand is expected to grow by about 27%, or 3,743 million tonnes oil equivalent (mtoe), worldwide from 2017 to 2040. According to IEA's New Policies Scenarios, carbon dioxide emissions from energy consumption would grow from 32.6 gigatons to 35.9 gigatons of carbon dioxide between 2017 and 2040, a rise of 10%. Whereas developed country carbon dioxide emissions are expected to drop by 23%, developing country emissions are anticipated to rise 27%.

- **Carbon Dioxide Emissions:** Under IEA's New Policies Scenario, carbon dioxide emissions from energy (including international bunker fuels) grow from 32.6 to 35.9 gigatons of carbon dioxide between 2017 and 2040, a rise of 10%. Whereas developed country carbon dioxide emissions are expected to drop by 23%, developing country emissions are anticipated to rise by 27%. The difference between developed and developing is starker when looking at power sector emissions. Developed countries are projected to decrease power sector emissions by about one-third by 2040 while developing countries are anticipated to see their emissions climb by about 20%.
- **Global Energy Demand Growth:** Energy demand is expected to grow by about 27%, or 3,743 million tonnes of oil equivalent (mtoe), worldwide from 2017 to 2040. Demand in developing countries—i.e., from the Organization of Economic Co-operation and Development (OECD) countries—are on course to shrink their combined energy demand by 4% (230 mtoe) in 2040. As a result, the share of global demand (less international bunker fuels) from developed countries falls from 36 to 30%. Developing countries, on the other hand, are on course to increase their combined energy demand by 45% (3,743 mtoe) and their share of global demand from 64 to 70%.

The International Energy Agency (IEA) just released its latest global energy forecast, the World Energy Outlook 2018 (*WEO2018*). Although IEA provides three different scenarios—the Current Policies, New Policies, and Sustainable Development Scenarios—we'll focus on its mid-range "New Policies Scenario," which includes current policies and expected or announced policies. IEA's forecasts peer 22 years into the future from an (estimated) 2017 baseline.

Here are the highlights:

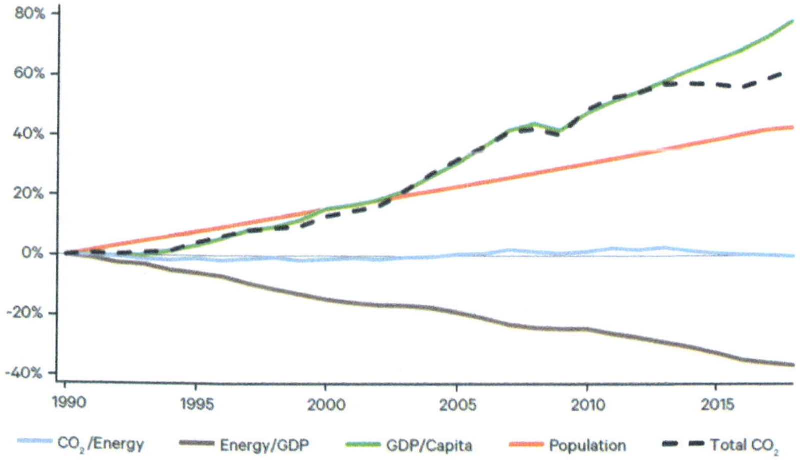

Acknowledgements The management of Covenant University is well-regarded for their unflinching publication support.

References

1. Dachowski R, Kostrzewa P (2016) The use of waste materials in the construction industry. Procedia Eng 161:754–758
2. Keskisaari A, Kärki T (2018) The use of waste materials in wood-plastic composites and their impact on the profitability of the product. Resour Conserv Recycl 134:257–261
3. Rose CM, Stegemann JA (2018) From waste management to component management in the construction industry. Sustainability 10(1):229
4. Davarpanah A, Razmjoo A, Mirshekari B (2018) An overview of management, recycling, and wasting disposal in the drilling operation of oil and gas wells in Iran. Cogent Environ Sci 4(1):1537066
5. Roy AS, Chingkheihunba A, Pakshirajan K (2016) An overview of production, properties, and uses of biodiesel from vegetable oil. In: Green fuels technology. Springer, Cham, pp 83–105
6. Hasanizadeh P, Moghimi H, Hamedi J (2017) Biosurfactant production by Mucor circinelloides on waste frying oil and possible uses in crude oil remediation. Water Sci Technol 76(7):1706–1714
7. Minx JC, Callaghan M, Lamb WF, Garard J, Edenhofer O (2017) Learning about climate change solutions in the IPCC and beyond. Environ Sci Policy 77:252–259
8. Oyedepo SO (2012) Efficient energy utilization as a tool for sustainable development in Nigeria. Int J Energy Environ Eng 3(1):11
9. Richard GN, Daniel R, Seth V, Brian P (2020) Global energy outlook 2020: energy transition or energy addition? With commentary on implications of the COVID-19 pandemic, Resources for the Future (RFF), Washington, DC, p 58
10. Olujobi OJ (2020) The legal sustainability of energy substitution in Nigeria's electric power sector: renewable energy as alternative. Prot Control Mod Power Syst 5(32):1–12. https://doi.org/10.1186/s41601-020-00179-3
11. Eleri EO, Ugwu O, Onuvae P (2012) Expanding access to pro-poor energy services in Nigeria. International Center for Energy, Environment and Development, Abuja, Nigeria, October 15
12. Mas'ud AA, Wirba AV, Muhammad-Sukki F, Mas'ud IA, Munir AB, Yunus NM (2015) An assessment of renewable energy readiness in Africa: case study of Nigeria and Cameroon. Renew Sustain Energy Rev 51:775–784
13. Wang Y, Xu L, Solangi YA (2020) Strategic renewable energy resources selection for Pakistan: based on SWOT-Fuzzy AHP approach. Sustain Cities Soc 52:101861
14. Akorede MF, Ibrahim O, Amuda SA, Otuoze AO, Olufeagba BJ (2017) Current status and outlook of renewable energy development in Nigeria. Niger J Technol 36(1):196–212
15. Stančin H, Mikulčić H, Wang X, Duić N (2020) A review on alternative fuels in future energy system. Renew Sustain Energy Rev 128:109927
16. Paulauskiene T, Bucas M, Laukinaite A (2019) Alternative fuels for marine applications: biomethanol-biodiesel-diesel blends. Fuel 248:161–167
17. Gabiña G, Martin L, Basurko OC, Clemente M, Aldekoa S, Uriondo Z (2019) Performance of marine diesel engine in propulsion mode with a waste oil-based alternative fuel. Fuel 235:259–268
18. Aakko-Saksa P, Koponen P, Roslund P, Laurikko J, Nylund NO, Karjalainen P, Rönkkö T, Timonen H (2020) Comprehensive emission characterisation of exhaust from alternative fuelled cars. Atmos Environ 236:117643
19. Sadeq AM, Bassiony MA, Elbashir AM, Ahmed SF, Khraisheh M (2019) Combustion and emissions of a diesel engine utilizing novel intake manifold designs and running on alternative fuels. Fuel 255:115769
20. Fayad MA, Tsolakis A, Martos FJ (2020) Influence of alternative fuels on combustion and characteristics of particulate matter morphology in a compression ignition diesel engine. Renew Energy 149:962–969
21. Dabi M (2020) Supervising UK. Implications of blended mesua ferrea linn oil on performance, combustion and emissions of compression ignition diesel engines. Therm Sci Eng Prog 19:100579

22. Atmanlı A, Ileri E, Yüksel B (2014) Experimental investigation of engine performance and exhaust emissions of a diesel engine fueled with diesel–n-butanol–vegetable oil blends. Energy Convers Manage 81:312–321
23. Yesilyurt MK, Aydin M, Yilbasi Z, Arslan M (2020) Investigation on the structural effects of the addition of alcohols having various chain lengths into the vegetable oil-biodiesel-diesel fuel blends: an attempt for improving the performance, combustion, and exhaust emission characteristics of a compression ignition engine. Fuel 269:117455
24. Agrawal BN, Sinha S, Kuzmin AV, Pinchuk VA (2019) Effect of vegetable oil share on combustion characteristics and thermal efficiency of diesel engine fueled with different blends. Therm Sci Eng Prog 14:100404
25. Sidibe S, Joel BL, Tizane DA, Vaitilingom G, Koulidiati J (2020) Comparative study of three ways of using Jatropha curcas vegetable oil in a direct injection diesel engine. Sci Afr 7:e00290
26. Yesilyurt MK, Aydin M (2020) Experimental investigation on the performance, combustion and exhaust emission characteristics of a compression-ignition engine fueled with cottonseed oil biodiesel/diethyl ether/diesel fuel blends. Energy Convers Manage 205:112355
27. Oyedepo SO, Airhihen PI, Adelekan DS, Ohijeagbon OI, Kelvin EU, Dirisu JO, Udo MO (2019) Experimental investigation of waste vegetable oil-diesel fuel blends effects on performance of compression ignition engine. Procedia Manufact 35:1039–1046
28. Agrawal BN, Sinha S, Singh DB, Bansal G (2020) Effects of blends of castor oil with pure diesel on performance parameters of direct injection compression ignition engine. Mater Today: Proc 21:1462–1466
29. Giridharan R, Rajan AV, Krishnan BR (2020). Performance and emission characteristics of algae oil in diesel engine. Mater Today: Proc
30. Hazar H, Sevinc H, Sap S (2019) Performance and emission properties of preheated and blended fennel vegetable oil in a coated diesel engine. Fuel 254:115677
31. Aydın S (2020) Comprehensive analysis of combustion, performance and emissions of power generator diesel engine fueled with different source of biodiesel blends. Energy 205:118074
32. Elżbieta D, Małgorzata S, Pośniak M, Andrzej S, Bartosz P, Woodburn J (2020) Exhaust emissions from diesel engines fuelled by different blends with the addition of nanomodifiers and HVO. Environ Pollut 259:113772
33. EL-Seesy AI, He Z, Hassan H, Balasubramanian D (2020) Improvement of combustion and emission characteristics of a diesel engine working with diesel/jojoba oil blends and butanol additive. Fuel 279:118433
34. Uslu S (2020) Optimization of diesel engine operating parameters fueled with palm oil-diesel blend: Comparative evaluation between response surface methodology (RSM) and artificial neural network (ANN). Fuel 276:117990
35. Murugapoopathi S, Vasudevan D (2019) Performance, combustion and emission characteristics on VCR multi-fuel engine running on methyl esters of rubber seed oil. J Therm Anal Calorim 138(2):1329–1343
36. Adaileh WM, AlQdah KS (2012) Performance of diesel engine fuelled by a biodiesel extracted from a waste cocking oil. Energy Procedia 18:1317–1334
37. Liaquat AM, Masjuki HH, Kalam MA, Varman M, Hazrat MA, Shahabuddin M, Mofijur M (2012) Application of blend fuels in a diesel engine. Energy Procedia 14:1124–1133
38. Sugözü İ, Cengiz ÖNER, Altun Ş (2010) The performance and emissions characteristics of a diesel engine fueled with biodiesel and diesel fuel. Uluslararası Mühendislik Araştırma ve Geliştirme Dergisi 2(1):50–53
39. Prakash R, Singh RK, Murugan S (2013) Utilization of biomass based fuel in a naturally aspirated diesel engine. Procedia Eng 51:501–507
40. Sharma S, Gupta S, Kumar N, Kumar S (2014) Comparative study on performance and emission characteristics of fish oil biodiesel and Mahua oil biodiesel blend with diesel and diesel fuel in a medium capacity compression ignition employing urea-SCR with Cu-ZSM5 (No. 2014-01-1499). SAE Technical Paper.
41. Sayin Kul B, Kahraman A (2016) Energy and exergy analyses of a diesel engine fuelled with biodiesel-diesel blends containing 5% bioethanol. Entropy 18(11):387

Fast Microwave-Assisted Pyrolysis of Wastes for Biofuels Production

Xin Jiat Lee

Abstract Non-renewable energies and renewable energies still remain to be imbalance up to today, with the over dependency on finite fossil fuels. The pros and cons of each energy technology have been discussed briefly, whereby among all those technologies, bioenergy shows great potential in the energy sector. To produce bioenergy, 2nd generation biofuels can be produced from waste materials of two main forms such as lignocellulosic and non-lignocellulosic origins, without competing with the food resources. To date, one of the most feasible thermochemical methods—pyrolysis can be set with fast heating rate and enhanced with microwave technology which is also known as fast microwave-assisted pyrolysis in order to produce high yield and high energy-dense biofuels with rapid processing. The fundamentals of pyrolysis and microwave technologies have been critically described along with their combination in operation setup for biofuel production. In this book chapter, various lignocellulosic and non-lignocellulosic wastes such as municipal wastes, agricultural wastes, plastics, and others have been reviewed as feedstock for fast microwave-assisted pyrolysis. Overall, this technology is viable however comes with the challenges in the scaling up stage and production issues. Nonetheless, with the current development, fast microwave-assisted pyrolysis can be a great form of waste-to-energy (WtE) or Energy from Waste (EfW) technology to be applied in the energy industry.

Keywords Waste · Pyrolysis · Microwave · Biofuel · Waste-to-wealth

---

X. J. Lee (✉)
Centre for Energy Sciences (ENERGY), Faculty of Engineering, University of Malaya, 50603 Kuala Lumpur, Malaysia
e-mail: leexinjiat@hotmail.my

Department of Mechanical Engineering, Faculty of Engineering, University of Malaya, 50603 Kuala Lumpur, Malaysia

A. Khan et al. (eds.), *Waste to Biofuel Technology*, Sustainable Materials and Technology, https://doi.org/10.1007/978-981-97-4561-6_5

1 Introduction

Energy is the essential necessity for domestic and industrial activities, to support the current development and future globalisation growth. Since the Industrial Revolution in the mid-eighteenth century, the energy demand has experienced an unprecedented surge from European and Western countries. From twentieth century onwards, Asia such as China, India, and other developing countries have also emerged as the key players in rapid industrialisation, with the need of great energy consumption to cater not only their growing populations, but also their industry development particularly for the manufacturing activities. Thus, over the last five decades, energy consumption has been increasing greatly at a global scale, to accommodate human population of 8 billions as of 2022. Generation of energy can be powered from both renewable and non-renewable sources. To date, more than half of the global power generation still depend heavily on conventional fossil fuels such as coal, oil, and natural gases, which are finite and non-renewable and create severe pollutions to the environment [1, 2]. Furthermore, despite the current advancement of fossil fuel-based energy generation with higher efficiency and lower pollutions, the emission of greenhouse gases (GHGs) such as carbon dioxide (CO_2), carbon monoxide (CO), nitrogen oxide (NO_X), and sulphur dioxide (SO_X), methane (CH_4) as well as other trace hydrocarbons to the environment is still inevitable and accumulating over the years [3]. Especially, at elevated levels against the background levels, GHGs are regarded as prominent air pollutants which can cause environmental impacts such as global warming due to their heat-trapping abilities, leading to higher sea level arisen from the melt of ice bergs and caps as well as extreme climate change around the world [4, 5].

In the endeavour of producing more sustainable and greener energies, biofuels are considered to be an attractive alternative as they are known to be green fuels with lower GHG emission especially CO_2, renewable, biodegradable with similar energy output comparable to fossil fuels [6, 7]. One of the most attractive features of biofuels is that almost any wastes can be converted into the feasible biofuels, whereby the synthesis path is sometimes referred as a waste-to-wealth (WtW) technology, making bioenergy to be sustainable and economic [8, 9]. Currently, there are numerous technologies available such as mechanical treatment, biological treatment, chemical treatment, and thermochemical treatment as well as any combination of treatments which can produce high-quality biofuels, depending on the feedstock's characteristics [10].

In recent years, generating biofuels from waste via thermochemical technologies such as combustion, co-firing, torrefaction, gasification, liquefaction, and pyrolysis are gaining more attention in the research community and industries due to their flexibilities and advancement to produce good quality products and by-products [11, 12]. Among those technologies, pyrolysis, as an established method, is widely researched due to its capability to generate three forms of biofuels such as biochar (solid fuel), bio-oil (liquid fuel) and bio-syngas (gas fuel), as feasible alternatives to conventional fossil fuels such as coal, petrol and gas [13]. As pyrolysis technologies progressing over the past 20 years, numerous forms of pyrolysis, from the conventional slow pyrolysis to the advanced technologies such as microwave-assisted pyrolysis, have been developed and thus, giving pyrolysis an edge over other thermochemical techniques [3]. More importantly, pyrolysis can handle almost any feedstock from non-organic to organic-rich wastes effectively with high impartiality. The derivation of biofuel from wastes, also known as 2nd generation biofuel through pyrolysis can aid the overwhelming landfill system as well as promote WtE/ EfW and WtW based on environmental and economic perspectives, respectively.

This chapter first analyses the current energy trends around the world and how bioenergy plays a crucial role in the future of energy development in comparison to other type of energies in Sect. 2. Following that, the fundamentals of pyrolysis as well as its advancement to fast pyrolysis system will be discussed in Sect. 3. The insight of microwave technology on pyrolysis will be conferred within this section as well, along with the know-how of the integration of microwave to pyrolysis process. For the following section, a critical discussion focuses on the available type of wastes will be discoursed, along the experimental studies which has applied the microwave-assisted pyrolysis, before concluding the future development and challenges.

2 Global Current and Future Energy Supply and Consumption Outlook

Based on the International Energy Agency (IEA) report, the global energy consumption has increased tremendously from 10,897.94 Terawatt-hour (TWh) in 1990 to 24,738.92 TWh in 2018, which is about 2.3 times of alarming surge over 28 years [14]. Developed and developing nations require enormous amount of energy for heavy industrial activities, transportation, agricultural activities as well as social and domestic activities, to promote technological and economic growths. From the IEA analysis investigation, a total of 9,337,702 kilo tonne of oil equivalent (ktoe) of energy was consumed in eight (8) main categories as shown in Fig. 1 [2]. As observed, both industrial and transportation activities contributed more than 50% of energy usage, up to 57.66%, followed by residential energy consumption (21.22%). Other lower contributors (<10%) include commercial and public services, non-energy application, agriculture activities, non-specific activities as well fishing industries. Thus, global energy applications to date, are mainly consumed by industries, transports, and domestics contributing over 78.88% globally.

To fully understand the energy sector outlook, the energy generation is equally important as well. From the current energy trend, to accommodate the consumption of power requirement, a total of 26,730.07 TWh of energy unit was produced against the usage of 24,738.92 TWh in surplus in 2018. As shown in Fig. 2, despite energy mix system has been adopted in the power generation, the use of fossil fuels such as coal (38.01%), oil (2.93%) and natural gas (23.01%) in combination (63.95%) still greatly exceeds the renewable energies, such as biofuels, waste, hydro, geothermal,

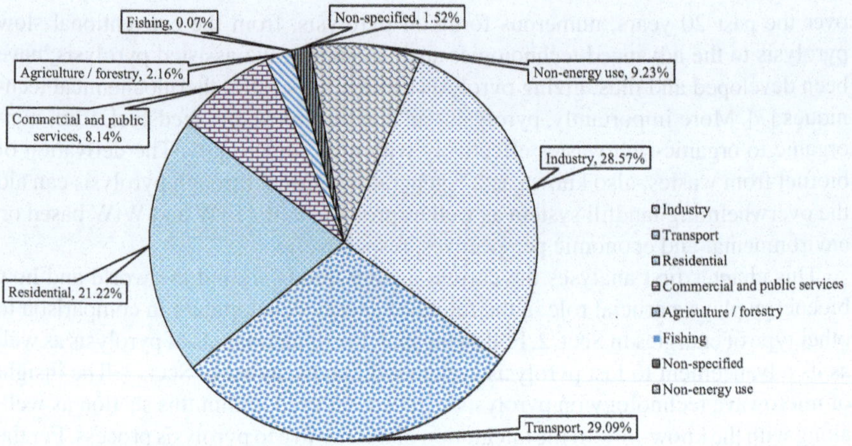

Fig. 1 Energy consumption based on different sectors in 2018 [2]

solar photovoltaic (PV), solar thermal, wind, tide and other forms. Unless a greater contribution from renewable energies, the price and usage of fossil fuel might likely to hike up an additional 20% in the coming 10 years [15]. Given such high dependency on fossil fuel, this will accelerate the deprivation of the fossil fuels in the near future, leading to possible adverse outcomes such as the collapse of energy sector in various countries as well as price volatility of fuel due to the imbalance of supply and demands. With the urgency to seek viable alternative energy supplies, renewable energies, especially, bioenergy should be focussed in the coming industrialisation and transportation technologies development, given that these are the major energy users globally.

2.1 *Non-renewable Energies*

Energy generation is mainly contributed by non-renewable energies and renewable energies, with the former constitutes over 50% of the global energy mix. Non-renewable energy, as indicated by its name, is a form of energy sourced from finite resources such as fossil fuels (natural gas, coal, and crude oil) and nuclear energy (uranium), which cannot be replenished in short period of time. Despite under the same non-renewable category, fossil fuel and nuclear energy are very different in terms of its generation mechanism as well as emission.

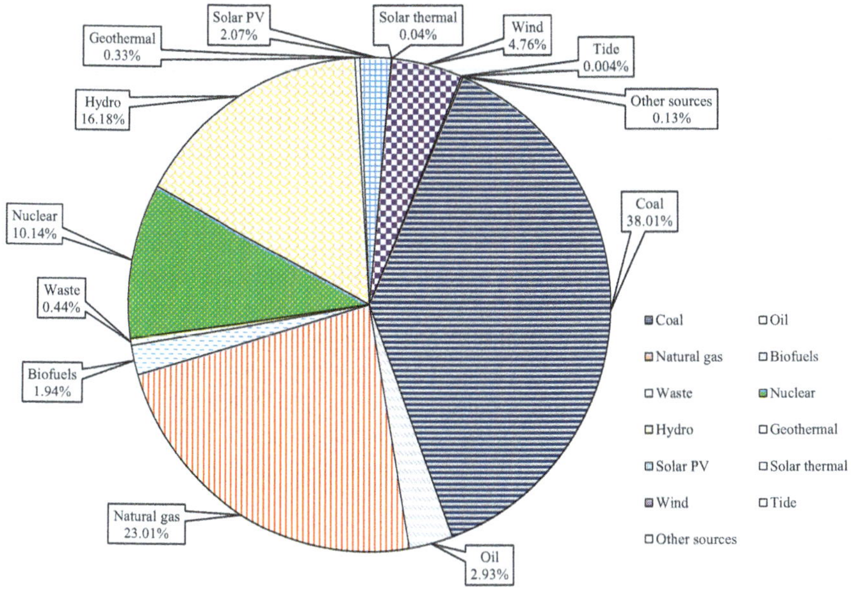

Fig. 2 Energy mix based on different sources in 2018 [16]

2.1.1 Fossil Fuels

Fossil fuels occurred naturally from decomposing of organisms under immense heat and pressure in the Earth's crust for hundreds of millions of years, which then formed into various energy products such as natural gas, coal, oil, petroleum, and tar sands [17]. For coal or coke, this form of fossil fuel is generated in swamps with the remains of dead organisms, which takes about 4–10 k years to accumulate into 1-m height of peat [18]. The grades of coal can be categorised into the highest grade being anthracite, followed by bituminous, subbituminous and the lowest grade of lignite coals [1]. Petroleum is mainly found in the ocean beds, and formed under an almost inert condition through bacterial breakdown. Relatively, petroleum combustion produces significantly lower carbon emission compared to coal burning [18]. Natural gases are highly pressurised products generated together with petroleum formation, which consist mainly of methane and ethane along with hydrocarbons such as ethylene, propane, hydrogen sulphide and others [17]. In some cases, natural gases can be found in their dry and crude form without the presence of petroleum, and these are known as non-associated petroleum gas. In terms of carbon emission, natural gases burning releases about 43 and 30% lower emission of CO_2, compared to coal and oil, respectively [18]. Exploration for oil and gas (O&G) reservoirs can be a costly operation and involves high capital expenditures (CAPEX). Once the reservoirs are found to be feasible for excavation, the drilling and refining facilities for O&G will be commissioned in the forms of oil rigs or offshore platforms, which can be in many types from man or unmanned, fixed or movable structures and others. All of which also require high operation expenditure (OPEX). As the O&G industry involves the handling of pressurised oil well with highly flammable and even explosive compounds, the offshore operation is high risk and further made more challenging with weather conditions, human factors and etc. Over 40 years of period, several notorious O&G incidents, including spill incidents and fire and explosion events, such as Piper Alpha accident 1988 [19], Exxon Valdez oil spillage 1989 [20] and Macondo well blow-out 2010 [21], have incurred severe environmental damages, huge economic losses and devastated fatalities. Furthermore, the instability of crude oil prices also impacts the job stability and operation of O&G facilities, whereby at times, crude oil prices can fluctuate more than 50% in a mere 6-month period [22]. While there has been an interest to shift energy generation from non-renewable to renewable and cleaner bioenergy, the current economic benefits of fossil fuel still outweighs in the global energy sector [23]. Nonetheless, it is expected that more bioenergy will be incorporated in the energy sector as bioenergy releases significantly lower amount of GHGs compared to fossil fuels.

2.1.2 Nuclear Energy

As a clean energy with a non-carbon generation process, nuclear energy generates vast amount of heat and energy through nuclear reactions, namely fission (splitting of nucleus) and fusion (combining of nuclei) mechanisms, with the former being the main contributor of power generation [24]. Therefore, with no combustion process taken place, nuclear energy basically generates zero carbon emission and almost zero or near-zero GHGs emission. Nonetheless, nuclear energy depends greatly on the finite main fuel source—uranium, a valuable rare earth mineral, under fission reaction can produce up to 200 Mega-electronvolt (MeV) of energy [24, 25]. Despite no waste carbon is emitted, gamma radiation emission and other radioactive wastes generation are inevitable, and can be detrimental to both environment and human health. From design to operation, each aspect associated with nuclear energy not only involves high technicalities, it also requires high discipline in process control and high CAPEX and OPEX. Due to its potential catastrophic impacts, nuclear plants are designed with the utmost safety features to minimise potential occurrence of accidents with fail-safe system [26]. Infamous incidents such as Chernobyl accident in 1986 due to engineering flaws and Fukishima Dai-ichi incident due to natural disasters, both highlight the potential catastrophic outcomes to human and environment in the event of accidents [27]. Furthermore, up to date, the disposal of radioactive wastes remains to be challenging and can only be resolved by prolonged storage until their half-life exhaustion [28]. Based on the data provided by International Atomic Energy Agency (IAEA), there are about 448 reactors in operation distributed over 32 countries, with United States of America (USA) having the most reactors (96 units), followed by France (58 units) and China (50 units) [29]. Currently, more countries are introducing nuclear powers into their power grid due to its massive generation and clean energy mechanism. Owing to the extravagant CAPEX and OPEX, commercialisation of nuclear energy especially in developing countries remains to be rather challenging.

2.2 *Renewable Energies*

In general definition, renewable energies are produced from sources which can replenish themselves naturally and continuously over time. Based on Fig. 2, renewable energies with the greatest portion is hydropower (16.18%), followed by wind (4.76%), solar PV (2.07%) and biofuels (1.94%), with the remaining renewable energies each having less than 1% contribution. While most renewable energies do not emit GHGs, their applications still exert minimal amount of carbon footprints, but significantly lower compared to non-renewable energies. Even with the current technological advancement, renewable energies are still underutilised as major countries like USA, China, and Russia are very much relying heavily on non-renewable energies with low adoption of renewable energies in their countries. Heavy dependencies on non-renewable energies should firstly be reduced gradually through energy-mix strategies with renewable energies and thereafter replaced completely at a planned manner, to avoid energy shortage.

2.2.1 Hydropower

Around 2000 years ago, hydropower started off from the invention of wooden waterwheel intended for milling grains [30]. Since then, hydropower has evolved from basic mechanical work to generating electricity for domestic use. In the early days, hydropower was very favourable for rural electrifications due to its simplicity in operation [31]. Theoretically, hydropower mainly gains the energy from the hydrokinetic force of flowing water impinging onto the turbines, with a wide range of hydro head heights such as high head (>50 m), medium head (10–50 m) and low head (<10 m) [30, 32, 33]. The flow from the dam's reservoir is discharged at controlled flowrate and angles to rotate the turbines at the most effective position [30, 31]. Up to this day, this form of renewable energy is still regarded as one of the most controversial technologies because massive lands and forests need to be cleared for dam construction, in order to create a large reservoir to retain the water at the upstream. Even though it is claimed that no GHG was released during operation, the pre-commission and commissioning of the dam usually involves invasive forest clearing, relocation of native communities, as well as land flooding which generates CO_2 eventually [32, 34]. However, in operation, dam construction at upstream actually serves very well in flood control and irrigation of water for downstream agriculture activities. To overcome the negative perception with hydropower, additional efforts such as tourism, employment creation, and local facilities development are undertaken at numerous dam operation sites [34, 35]. Three Gorges Dam in China (operating capacity of 22,550 MW) and Bakun Dam in Malaysia (operating capacity of 2400 MW) are infamous examples which involved forest clearing, major population relocation, and irreversible ecological impact due to drastic geographical change [36–38].

As mega dam for hydropower involves high CAPEX, complex design as well as political and public supports, small-scale hydropower (10 MW max) including mini-hydro (<2 MW), micro-hydro (<500 KW) and pico-hydro (<10 KW), are more environmentally-benign as they requires no dam construction to generate electricity [30]. By only diverting one part or more parts of the moving water stream from the main river stream via a barrier called weir, the "run-of-river" will then impinge the flow through the turbine system to generate energy in return [39]. Countries with extensive river streams such as China, Japan, USA and India are among the countries with good progress with this technology [40]. Even though with advantages such as lower investment cost, simpler setup and maintenance (low OPEX), long operation period (up to 50 years), small-scale hydropower however, is site-specific and can be affected by seasonal flow change and impose certain impacts on the fisheries and agricultural industries [30].

2.2.2 Wind Energy

By concept, wind refers to the air motion arisen from the differential thermal conditions between lands, oceans, or both, and over the world, wind is available continuously day and night with the power potential of 10 million MW [41]. Historically, the use of wind started 3000 years ago by converting the wind motion as kinetic energy for pumping water, grinding grains as well as sailing boats and ships [42]. Nonetheless, the first wind turbine was only built in Scotland on 1887 years for energy generation, and was further developed to medium-sized (300 kW) wind turbine in the early 2000 [43]. With gaining interest from highly populated countries such as India, China and Australia after 2000s to seek more sustainable energy generation, the wind turbine operating capacities was slowly scaled up to 500–600 kW [42]. Now, in modern wind energy sector, the current capacities of the wind turbines can reach in between 1.5 and 2.5 MW [44]. Wind power gain is directly proportional to the 3rd of the wind speed [44]. In the current practice, wind power is harnessed through a collective wind turbines on farms which can be stationed offshore or onshore, and at high-altitude sites with stronger and consistent wind turbulence [45]. Major players of wind power are mainly focused at Western countries such as Germany, Spain, Denmark and USA, in the form of wind farms [46]. While for South and Central of America, the renewable energy policy is still weak and comes with price issues which hinder the development of wind energy locally despite blessed with abundance of wind resources [42]. Similarly in Asia countries, wind energy is not gaining popularity at the moment as well due to internal factors such as high CAPEX and land area issues [47]. Based on the current market, the wind energy is still regarded to be clean, renewable and economic in term of OPEX; but, its site selection and construction of wind farm comes with several challenges. For instance, wind motion or distribution in current technology can be accurately detected and even predicted of its direction and speed. Certain degree of uncertainties however are unavoidable as the nature of wind can alter in the short period of time causing inconsistency or non-persistence for long operation [44]. Even faced with great challenges, these challenges can be overcome by focusing more on the turbine design development to be stronger, lighter and more adaptable to the change of wind speed and direction, for more efficient electricity generation [46]. Overall, the harness process from wind motion is greatly influenced by wind distribution, geographical suitability and available land for massive installation of wind turbines.

2.2.3 Solar Energy

The sun is a great source of unlimited energy and about 1368 W m^{-2} of its radiation can reach the Earth's outer atmosphere all day long which can be tapped for energy generation [48]. To harness the power of sun, there are basically two fundamental ways of harnessing, namely by solar photovoltaic (PV) or solar thermal or combination systems. Solar PV utilises a device to which can absorb the lights from the sun and then covert the photon directly into electricity [49]. The materials used for PV synthesis varies but the most common and economic to date is still silicon-based cell which result in different forms of cell produced from monofacial to bifacial solar PVs [50]. To improve the low conversion of conventional PV cells, more advanced materials such as thin-film solar cells made from semiconductor-based elements which include zinc telluride, cadmium telluride, cadmium-nickel doped materials [51, 52] as well as Group III and Group V element (III–V) solar cells produced from gallium, indium and antimony [53, 54], have been researched recently. Aforementioned, conventional solar PV suffers low conversion efficiency, and thus, many materials have been attempted to solve this issue. Up to date, perovskites solar cells have shown great potential in photon-based electricity generation, which great efforts were observed in the research communities in recent years [55].

Solar thermal system works by collecting and intensifying the sunlight to harness the heat for electricity generation, and comes with different capacities of collectors such as low- (heating of small reservoir), medium- (heating water/air for domestic and commercial uses), and high-temperature (electricity generation) collectors depending on the applications [56]. In working principle, a series of mirrors/reflectors will be used to direct and focus the sunlight onto a system containing heat-transfer medium which then be utilised for electricity production. Currently, Germany, as one of the largest solar PV producers, can generate an average solar irradiance of 1000 kWh m^{-2} [41]. To further enhance the solar energy harness, combination of quantum and thermal system into solar PV not only recover the waste heat from solar converter, this system also greatly increases the overall electricity generation exceeding 50% [57]. From the market trend, solar energy is gaining momentum in Asia, especially China, leading to the great reduction of PV price. Subsequently, the CAPEX for this technology will be more affordable to developing countries. Once installed, the maintenance and operation of solar power should be at its minimum, giving an overall lower OPEX. With the progressing of technology, solar energy will likely to increase its portion in energy mix in the future.

2.2.4 Tidal Energy

The gravitational and centrifugal forces between the Earth and cosmic attractions from the moon and sun, can create the tide motion with great potential energy up to 2.5 TW [58]. Thus, on the ocean, it is said that due to closer distance with Earth, the attraction from the moon is stronger compared to the sun [59]. Tidal energy, in the forms of potential energy or kinetic energy, can also be extracted via tidal barrage or tidal current turbine. Tidal barrages create an ocean basin to collect sufficient head before emptying the basin in a repetition cycle to generate electricity [60]. To date, tidal barrage is a commercialised system that comes with different designs such as ebb generation, floor generation, two-way generation and double-basin tidal barrages. France (480 GWh per year), Canada (30 GWh per year) and Russia (21.8 GWh per year) are utilising this technology in their energy mix [59]. However, despite the good and clean conversion to electricity, this method involves the construction of foreign structure which can withstand the changing tides, often requires high construction cost and the structure will directly impact on the marine life. Overall, tidal barrage can be high in CAPEX and OPEX, and incur certain environmental impact to the shore areas.

On the other handm tidal current turbine is similar to wind turbine by extracting the kinetic energy from the motion of waves [61]. Typically, the turbine will be built under the seabed at least 1 km or more away from the shore, with the water depths in between 20 and 30 m [62]. Surprisingly, the prediction for tidal conditions such as the speed and direction are relatively more accurately compared to wind conditions, which also means that the power generation via tidal waves could be more easily planned and optimised in the long run [62]. More popular in United Kingdom (UK), USA and Norway, tidal energy is harvested at their shoreline as well as offshore to optimise the electricity generation [62]. Despite being a renewable energy, the utilisation of tidal energy faces great challenges technologically such as maintenance difficulty and installation of grid connection translating to high OPEX, along with great CAPEX as well as direct environmental and aesthetic impacts [59]. Nevertheless, from an overview, the commercialisation of tidal energy has been progressing well globally.

2.2.5 Geothermal Energy

Geothermal or heat from the ground has been in the Earth's crust and generated from the decomposition of the radioactive isotopes of uranium (^{235}U and ^{238}U) and other minor isotopes [63, 64]. However, not many countries (around 80 countries) possess this form of renewable energy, which largely depends on their geographical location [65]. Normally, geothermal energies existed in underground streams, and at times, those energy sources can reach to the surface and form hot pool or hot spring. However, for underground sources, the extraction requires drilling and welling processes, which can form two types of wells depending on their temperatures range, namely high-temperature well (>150 °C) and low-temperature well

(<150 °C) [63]. So far, geothermal power has been directly for electricity generation as well as to power heat pumps and heat spaces, swimming pools, bathing and recreational purposes without any issues [66, 67]. Since the heating from Earth's Crust is available for almost 24 h without intermittent issues, the recovery of enthalpy from geothermal fluids can generate consistent electricity [45, 68]. However, the use of geothermal energy comes with shortfalls such as high exploration cost from drilling and welling (high CAPEX) as well as the possible contamination by traces of ammonia, mercury and other toxic elements [34].

2.2.6 Wastes Energy

Wastes generally refer to domestic wastes or municipal solid wastes (MSWs) produced from human activities such as food wastes, garden wastes, plastics wastes, metals, glasses and papers as well as mixture of semi-solid and even liquid wastes [41]. With the growing population globally, MSWs are also rapidly being generated with an annual rate of 3.2–4.5% for developed nations and 2.0–3.0% for developing countries, respectively [69]. Thus, in weight basis, about 0.5–0.8 kg of MSWs are produced for one person, but up to 1.7 kg of MSWs for those who live in city areas [48, 69]. Therefore, waste energy should be utilised not just to generate power, but also to reduce the overwhelming landfill.

Basically to produce electricity from waste, MSWs are either directly incinerated for power generation or converted into combustible biofuels (bioenergy) via thermochemical methods and biological treatments. This form of practice is known as WtE or EfW. Nonetheless, not all MSWs are combustible, in this case, waste materials such as metals and glasses can be segregated for recycling and other treatments. Therefore, by excluding those non-combustible materials, MSWs typically contain about 75% of combustible factions, generating up to 10,000 kJ kg^{-1} if recovered [35]. Of course, apart from incineration or thermochemical treatments, WtE also converts MSWs which are rich with organic portions to ethanol via fermentation, or to landfill biogas (CO_2 and CH_4) via anaerobic digestion [35]. In some WtE facilities, they are readily equipped with CH_4 collection and recovery systems directly used for electricity generation, thereby directly eliminating the emission of GHG (~23 times greater in heat-trapping compared to CO_2) [41]. Incineration of wastes is only largely practised in developed nations such as Japan and USA as the cost of technology setup and maintenance is high (high CAPEX and OPEX) and associated with complex operation from collection, segregation, combustion and residual waste (ash) disposal. During the incineration process, harmful emissions such toxic air pollutants (dioxins furans) and acid gases are generated and required to be treatment before discharging to the environment. Over the years, this technology has matured into mega-industrial scale capable of handling thousands of tonnes of wastes per day despite facing challenges especially at the waste segregation process [69].

2.2.7 Bioenergy

The current biofuels in the energy mix still remain low (~1.94%), but bioenergies have been gaining more shares in the global energy sector with about 5–6% annual increment against the past years [16]. Depending on the origin of the feedstock, some of the biofuels can also be part of waste energy if they are made from waste materials such as MSWs, agricultural by-products and industrial wastes. However, currently, most biofuels are in the forms of bioethanol, biodiesel and hydrotreated vegetable oil (HVO), solid biofuels and biogases, mainly derived from edible oils and food and animal fats [70, 71]. Despite biofuels emit lesser GHGs emit lesser GHGs, as well as carcinogens and sulphur particulates [72], bioenergy still faces controversial debate for energy over food crisis given that famine issue remains to be one of critical issues globally. Furthermore, edible feedstock for biofuels production also contributes similar carbon footprint amounts as fossil fuels in terms of non-GHG related impact categories from life cycle analysis (LCA), rendering it to be less sustainable [73]. To resolve the energy over food issue, emerging researches now focus on the exploration of non-food feedstock, such as wastes materials including lignocellulosic and non-lignocellulosic biomass [74].

As a way to standardise the global development of biofuels, biofuels are generalised based on the nature of the feedstock used as follows: 1st Generation biofuel (1st Gen), 2nd Generation biofuel (2nd Gen) and 3rd Generation biofuel (3rd Gen). For 1st Gen, this started with the use of edible feedstock such as vegetable oil, sugar and other organic compounds, by applying biochemical processes such as fermentation and transesterification. Fermentation involves the conversion of sugar or starch into bioethanol whereas for transesterification, vegetable oils or animal fats under the aid of catalysts are converted into biodiesel. To date, numerous commercialised biodiesel products have been successfully developed and used in blending with 95% of petroleum for vehicle application [40]. However, the issues associated with food security, competition with land area for cultivation as well as limited supply have hindered 1st Gen development in energy sector. Therefore, recently, 2nd Gen and 3rd Gen have emerged as the next frontier in bioenergy. Primarily, 2nd Gen only uses mainly wastes, renewable biomass and non-edible oils as feedstock for biofuels generation. Biomass can be in many forms, from non-living and living materials, from lignocellulosic and non-lignocellulosic as well as from domestic to industrial discharges [13]. The varieties of biomass such as oil palm wastes [6, 74], spent coffee ground [75], MSW sludge [9], corn cob [76], and others can be converted into different biofuels suitable for heat and electricity production, without competing with the food resources [77]. Non-edible oils such as jatropha, castor, karanja, rubber seed and sea mango also can be sourced as the precursors for biodiesel synthesis [40]. However, despite being non-competition with food crops, 2nd Gen still requires agriculture lands for cultivations. Therefore, 3rd Gen comes into the picture, whereby engineered algae cultivation only requires non-arable lands, possesses high CO_2 adsorption with great mass generation as well rapid growth rate [3]. However, 3rd Gen despite has been extensively researched, their applications in industrial scale are still under development due to the technological restrictions on cultivation of algae and their costly post-processing treatment to dewater the algae. Recently, there has been an interest in applying co-feedstock with non-renewable portion (i.e. waste plastics, scum and others) to produce higher calorific biofuels [3, 78]. This form of biofuel would still be categorised as 2nd Gen/3rd Gen as determined by the major portion of the feedstock; however, this type of biofuel would not be claimed as fully renewable in this case. In comparison, full biomass-based energy provides more diverse and renewable energy resources, with a net emission of GHGs [79]. Given its full potential with high varieties in feedstock, bioenergy can provide massive energy supplies annually with the current biomass supplies [80]. Globally, most countries are also trying to push bioenergy into their energy grid in order to reduce the dependency of the finite fossil fuel.

2.3 *Overview of Energy Sector and Role of Bioenergy*

As of 2020, IEA and Nuclear Energy Agency (NEA) collaborated and analysed both Organisation for Economic Co-operation and Development Nuclear Energy Agency (OECD) and Non-OECD countries, to produce a levelised cost of electricity (LCOE) for comparison purpose, and the unit price for both non-renewable and renewable energies at the discount rate of 7% are summarised in Table 1. However, the data should be conservatively referenced as not all countries (only 23 countries with 243 power plants) submitted their electricity costs for 2020 evaluation with no availability of ocean energy data in this cycle. Nonetheless, based on previous literature [81], ocean energy varied in between 18 and 24 USD MWh^{-1} in 2011, and expected to be lower in 2020 with further technological advancement. As observed, the prices of non-renewable energies (32–116 USD MWh^{-1}) and renewable energies (50–126 USD MWh^{-1}) are comparable (excluding ocean energy), and in fact, replacing fossil fuels would not pose any economic risks since the prices for renewable energies are getting lower over the years. To reduce the impacts of GHGs especially CO_2, fossil fuel processing facilities are gradually introduced with carbon capture and storage (CCS) system. However, from the pricing wise, CCS has greatly increased the unit price of coal (31.82% increase), lignite (22.11% increase) and gas (28.17% increase) energies. Among the renewable energies, wind energy facilities at onshore produced the lowest cost (50 USD MWh^{-1}) while solar PV (residential) remained to be the highest cost (126 USD MWh^{-1}), which could be explained by the fact that the domestic conversion was not as efficient as commercial and utility scale systems. In bioenergy, biomass energy cost (118 USD MWh^{-1}) was about 34.09% more expensive compared to coal energy. However, compared against coal energy (CCS), it was merely 1.72% difference in their unit prices. Thus, to have a cleaner energy production from coal, CCS despite can achieve this objective but comes with a great increase in price. Therefore, biomass can be upgraded into more energy efficient and cleaner biofuels via thermochemical techniques (pyrolysis, torrefaction, etc.), which will have comparable energy contents as coal with lower emission of GHGs in a long run [13].

Table 1 Comparison of energies in terms of levelised cost of electricity cost [82]

Technology	Median levelised cost of electricity (USD MWh^{-1})
Non-renewable energies[a]	
Coal	75–110 (Median: 88)
Coal (CCS)	113–119 (Median: 116)
Lignite	93–97 (Median: 95)
Lignite (CCS)	116–116 (Median: 116)
Gas (CCGT)	42–107 (Median: 71)
Gas (CCGT, CCS)	70–112 (Median: 91)
Nuclear	42–102 (Median: 69)
Nuclear (LTO)	30–35 (Median: 32)
Renewable energies	
Hydro (reservoir, >=5 MW)	39–142 (Median: 72)
Hydro (run-of-river, >=5 MW)	46–104 (Median: 68)
Wind onshore (>=1 MW)	29–140 (Median: 50)
Wind offshore	49–200 (Median: 88)
Solar PV (utility scale)	34–172 (Median: 56)
Solar PV (commercial)	74–140 (Median: 94)
Solar PV (residential)	108–223 (Median: 126)
Solar thermal (concentrated solar power)	112–130 (Median: 121)
Geothermal	78–120 (Median: 99)
Biomass	53–182 (Median: 118)
Ocean[b]	18–24 (Median: -)

[a]Abbreviation: CCS = Carbon capture and storage; CCGT = Combined cycle gas turbine; LTO = long-term operation
[b]Data extracted from Bruckner et al. [81]

To quantify the environmental impacts of energies production, LCA, an established cradle-to-grave analysis which investigates a process/a product/an activity from the origin of the raw materials to the end of product/process lifespan, has been more comprehensively studied on both non-renewable [83] and renewable energies [84–87], over a 20-year time frame. In energy analysis, the overall emission per unit energy also signifies how green the energies used and Table 2 also provides the life cycle CO_2 equivalent (CO_{2eq}) emission including albedo effect as studied by Intergovernmental Panel on Climate Change (IPCC) [88]. As observed from Table 2, coal and gas energy generation release 820 g CO_2 kWh^{-1} and 490 g CO_2 kWh^{-1}, respectively, generally higher in comparison to other renewable energies [34]. Nuclear energy (12 g CO_2 kWh^{-1}) and wind energy (11–12 g CO_2 kWh^{-1}) are among the greenest technologies with their low GHGs emission. Bioenergy (230–740 g CO_2 kWh^{-1}) emitted generally lower GHGs amounts compared to coal power generation. With the current development with bioenergy, it is anticipated that its unit price and

Table 2 Comparison of energies in terms life cycle emission [88]

Technology	Life cycle emission (g CO_{2eq} kWh^{-1})
Non-renewable energies	
Coal (Pulverised)	740–910 (Median: 820)
Gas (CCGT)	410–650 (Median: 490)
Nuclear	3.7–110 (Median: 12)
Renewable energies	
Hydro	1–2200 (Median: 24)
Wind onshore	7–56 (Median: 11)
Wind offshore	8–35 (Median: 12)
Solar PV (rooftop)	26–60 (Median: 41)
Solar PV (utility)	18–180 (Median: 48)
Solar thermal (concentrated solar power)	8.8–63 (Median: 27)
Geothermal	6–79 (Median: 38)
Biomass (co-firing)	620–890 (Median: 740)
Biomass (dedicated)	130–420 (Median: 230)
Ocean	5.6–28 (Median: 17)

carbon emission will reduce over time and likely to be the focal the research and energy sector for coming years, especially with the application of pyrolysis-based technologies.

3 Pyrolysis Technology

Before technological development in the older days, wood logs were converted into charcoal through prolonged burning for days and sometimes up to a month in airtight kiln, emitting black smoke during the process. Little did people know back then that this process can be adopted into one of the thermochemical processes which is clean, efficient and fast, namely pyrolysis in the modern days. Today, this form of technology can generate three phases of biofuel products such as solid products (i.e.: charcoal, biochar, bio-charcoal and tar), liquid products (i.e.: pyrolysis oil, bio-oil and bio-crude) and gas products (syngas and bio-syngas), under precise parameters control (i.e.: heating rate, temperature, dwelling time, particle size as well as inert type and flowrate) to maximise the desired yield portions of the products.

With the rapid development on thermochemical technologies, pyrolysis has grown into different branches, namely conventional and advanced pyrolysis [3]. Conventional technique mainly involves the use of single pyrolysis technology but by varying the heating rates, this results in slow, fast and flash pyrolysis. For advanced technique, these technologies can be differentiated by introducing the use of other technologies as an enhancement to the existing conventional process, such as the aid of catalyst (catalytic pyrolysis), mixture of feedstock (co-pyrolysis), hydrogen-based (H_2-based) technology (hydropyrolysis) and last but not least, microwave technology (microwave-assisted pyrolysis). Several reviews have critically studied the current pyrolysis technologies as well as their advancement in recent years [3, 9, 10, 89]. Therefore, in this book section, the focus will be placed mainly on microwave-pyrolysis studies. In this section, the scope will start to discuss the fast pyrolysis technology and mechanism, and followed by microwave fundamental, and finally focusing on fast microwave-assisted pyrolysis technology on different waste feedstock.

3.1 Fast Pyrolysis Technology and Mechanism

General modes of pyrolysis include slow, fast and flash pyrolysis are mainly governed by their heating rates with the following parametric settings [11, 13, 90]:

- Slow pyrolysis (heating rate <60 °C min^{-1}; pyrolysis temperature: 550–1000 °C; residence time >5 min)
- Fast pyrolysis (heating rate >60 °C min^{-1}; pyrolysis temperature: 800–1200 °C; residence time: 0.5–10 s)
- Flash pyrolysis (heating rate >1000 °C s^{-1}; pyrolysis temperature: 700–1200 °C; residence time <2 s)

Based on the pyrolysis setting, fast pyrolysis process shows a fast heating rate at elevated temperature with a short residence time, which in favour of the bio-oil formation [74]. This differs greatly from slow pyrolysis, as this form of pyrolysis drives the formation of biochar as the major fraction among the pyrolytic products formed [6]. Thus, it is imperative to understand the underlying mechanism of pyrolysis on different biochemical contents, such as cellulose, hemicellulose, lignin, protein, water, ash and carbohydrate depending on the wastes' characteristics [91].

3.1.1 General Mechanisms of Pyrolysis on Biochemical Contents of Wastes

Waste materials can be in many forms, with different compositions. For instance, waste biomass can be lignocellulosic and non-lignocellulosic nature, which can affect the biofuels' yields and characteristics significantly. Nonetheless, in the general mechanism, formation of pyrolytic products from wastes can be generated through the following steps [13]:

$$\text{Waste} \rightarrow \text{Water} + \text{Unreacted portion} \tag{1}$$

$$\text{Unreacted portion} \rightarrow (\text{Gas Portion} + \text{Volatile})_1 + \text{Solid Portion}_1 \quad (2)$$

$$\text{Solid Portion}_1 \rightarrow (\text{Gas Portion} + \text{Volatile})_2 + \text{Solid Portion}_2 + \text{Tar} \quad (3)$$

For fast pyrolysis, Eqs. (1–3) are equally important for maximise bio-oil formation. When the waste materials are firstly subjected to thermal pyrolysis, they would degrade into unreacted portion by releasing the water as moisture. Further thermal decomposition would convert the unreacted portions into two main components—main products (gas portion_1 and volatile_1) and intermediate product (solid portion_1). Under longer residence time, solid portion_1 would undergo the secondary reaction to form majority of gas portion_2 and minority of volatile_2 as well as the undesirable tar product. To inhibit the formation of biochar and gas portions through secondary reaction, accurate pyrolysis parametric control is needed to balance the fast heating rate to favour the generation of volatile, without promoting the formation of gas portion at elevated temperature. The following section details the thermal degradation of biochemical contents

3.1.2 Pyrolysis Mechanistic Steps on Biochemical Contents of Wastes

Table 3 proposes the thermal decomposition of biochemicals based on different temperature zonation for pyrolysis mechanistic steps. However, the temperature zonation should be referred as a general reference, as it is understood that waste materials are complex in nature, with varying biochemical contents; thereby affecting their overall rate and temperature range of decompositions. Generally, starting from 100 °C, water in the wastes materials are assumed to be dried off for lignocellulosic-based materials in Zone I. This is not the case to non-lignocellulosic materials for some which are moisture absorbing, they would need longer drying time at higher temperature (Zone A).

Typically, extractives are available on lignocellulosic-based materials such as pigments, wax and other minor traces compound, and would need slightly higher temperature for thermal degradation (up to 250 °C) (Zone II). When above 250 °C, the hemicellulose which is less thermal stable compared to cellulose, starts to degrade [92]. In Zone III, hemicellulose would undergo dehydration and depolymerisation reactions to generate various products such as acids, chars, waters, levoglucosan, furans and others [8, 93]. For cellulose which is slightly recalcitrant towards thermal effect, these compounds would only experience dehydration when the temperature is below 350 °C (Zone IV). However, when the thermal process progresses to 450 °C, depolymerisation of cellulose is dominant in this case up to 600 °C whereby fragmentation would occur in Zone V [94, 95]. Among those biochemical contents, lignin is the most resistance against thermal degradation up to 500 °C [93]. Overall, the stability of these three components against thermal treatment is ranked as follows: lignin > cellulose > hemicellulose with a reducing trend.

Table 3 Temperature zonation scheme on biochemical contents of wastes materials

Lignocellulosic pyrolysis scheme[a]	Non-lignocellulosic pyrolysis scheme[b]
• Zone I: 100 °C—Drying of moisture removal	• Zone A: 100–200 °C—Moisture removal and evolution
• Zone II: 100–250 °C—Elimination of extractives	• Zone B: 200–350 °C—Breakdown of protein, carbohydrate and decomposition
• Zone III: 250–350 °C—Hemicellulose degradation and reaction	• Zone C: 350–550—Degradation of lipid and devolatilisation of protein and carbohydrate
• Zone IV: 250–500 °C—Cellulose and lignin decomposition and reaction	• Zone D: >550 °C—Vaporisation of non-volatile components
• Zone V: > 500 °C—Lignin decomposition and reaction	

[a] Adopted from Raveendran et al. [96]
[b] Adopted from Lee et al. (2020)

Non-lignocellulosic materials and wastes thermal degradation behaviour can be completely different, with main biochemical contents such as protein, carbohydrate and lipid [97]. As a starter of the thermal process, moisture removal for non-lignocellulosic materials and wastes will be in Zone A (100–200 °C) which has a wider temperature range compared to lignocellulosic-based materials and wastes. It is known that multiple thermal degradation reactions would occur individually and concurrently for protein and carbohydrate in the Zone B. However for lipid, its predominant degradation zone would be in Zone C. When reaching above 550 °C (Zone D), some of the non-volatile materials might be converted into gaseous portion. For bio-oil formation, lipid and protein are more readily to be converted into liquid portion of the biofuels as compared to lipid [3].

3.2 *Microwave Technology*

3.2.1 Fundamental and Mechanism

Microwave technology has first gained major breakthrough in food industry for culinary purpose and since then, it popularises into the homes of many people as microwave ovens. Other than the use for cooking, microwave also has been used in wastewater treatment [98], drying technology [99], communication [100] and others. The superiority and suitability of microwave technology only gain notice in pyrolysis for a mere 10–15 years ago [101, 102]. Fundamentally, grouped as one of the electromagnetic spectra, microwave is a form electromagnetic irradiation with the wavelength in between 1 mm and 1 m, corresponding to the frequency in between 300 GHz and 300 MHz. Despite the wide range of working wavelengths, all microwave-based devices have been operating a fixed frequency (2.45 GHz) and wavelength (0.1225 m) as well as photon energy (1.6 e^{-3} eV) to ensure non-disruption to other microwave applications [103].

To generate heat via microwave, dipolar polarisation mainly contributes to this purpose aside ionic migration, through two forms of perpendicular components such as electric and magnetic fields [104]. For microwave dielectrics, they are known to effectively absorb microwave which are also excellent materials for microwave heating compared to insulator and conductor [103]. Thus, before running microwave-assisted pyrolysis, the feedstock should be assessed of its microwave absorbability or dielectric properties through the property known as loss tangent (tan δ), which is the ratio of the dielectric loss (ε'') to dielectric constant (ε') [101]. Basically, materials can be grouped as follows according to their tan δ values:

- tan $\delta < 0.1$ represents low microwave absorbability;
- $0.1 <$ tan $\delta < 0.5$ represent intermediate microwave absorbability
- tan $\delta > 0.5$ represents high microwave absorbability.

Microwave heating also favours the presence of water/moisture (tan $\delta = 0.123$), which also explains why dewatering of feedstock is usually not required in microwave-assisted pyrolysis [101]. With higher values of tan δ, this also implies that better wave absorbability the material possesses leading to better heat transfer. From Table 4, a comprehensive list of tan δ values is recorded accordingly, which can be an excellent reference for researchers to select the good dielectric materials.

3.2.2 Comparison of Microwave Heating and Conventional Heating

Microwave heating and conventional pyrolysis heating are illustrated in Fig. 3. As shown for conventional pyrolysis heating, it usually involves the use of electrical furnace, which takes time to generate the heat for thermal degradation of the waste solids. As the heat becomes more intense, a temperature gradient is formed between the outer surfaces of the feedstock and the heat zone in the furnace, thereby driving the heat exchange through surface transfer to the inner core of the waste solids via conduction, convection and radiation [106]. However, heat conductance or thermal conductivity of the waste materials can be weak, leading to the uneven and slow heat-transfer process.

Microwave heating, however, is a form of dielectric heating which effectively transfers the electromagnetic power and penetrates into the core of the waste materials directly [103]. Therefore, on the contrary to conventional heating, the inner core shows higher temperature, creating a thermal gradient from inner to outer of the waste materials. Therefore, heating of wastes via microwave energy is more rapid, efficient and targeted in this case, without the need for agitation or for fluidisation [107]. Overall, microwave heating is specific with rapid conversion of energy to heating, reducing the pyrolysis time and completion of reaction time as well.

For conventional heating method, the high moisture or wet feedstock is not desirable and needs to be reduced to 10%; else, the bio-oil quality will deteriorate [13, 109]. Moreover, conventional heating also has a problem in handling large waste

Table 4 tan δ values at 2.45 GHz [101, 104–106]

Material/feedstock	tan δ (dimensionless)
Lignocellulosic materials biomass	
Aspen bark	0.22
Pine wood	0.19
Pine bark	0.18
Particle board	0.1–1.0
Fir plywood	0.01–0.05
Wood polymer	0.03
Rubber	0.002–0.005
Oil palm shell	0.12
Oil palm fibre	0.08
Oil palm empty fruit bunch	0.297–0.535
Sludge	0.035
Acid washed sludge	0.019
Carbon materials	
Carbon black	0.23
Graphite powder	0.36–0.67
Graphite carbon powder	0.40–1.0
Empty fruit bunch biochar	0.134
Charcoal	0.14–0.38
Coconut activated carbon	1.646
Activated carbon	0.31–0.90
Chemical and biochemical contents	
Lipid/fat	0.105–0.424
Carbohydrate/cellulose	0.035
Protein	0.068
Water	0.123
Lignin	0.052
Hemicellulose	0.062
Minerals and metallic compounds	
ZnO	0.03
TiO_2	0.002–0.005
Fe_2O_3	0.001–0.05
Al_2O_3	0.0006–0.0007
MgO	0.0004–0.0005
Silicon	<0.012
SiO_2	0.0002–0.0003
SiC	0.37
Polymer	

(continued)

Table 4 (continued)

Material/feedstock	tan δ (dimensionless)
Polypropylene	0.0003–0.0004
Polystyrene	0.0002–0.0003
Polyethylene	0.0001–0.0003
Polyvinyl chloride	0.0056
Teflon Perfluoroalkoxy (PFA)	0.00015–0.00048
Glass and ceramics	
Borosilicate glass	0.00106–0.00120
Plexiglass	0.0057
Fused quartz	0.00003–0.00006
Ceramic F-66	0.00055
Porcelain No 4462	0.0011
Organic compounds	
Ethanol	0.941
Methanol	0.659
1-butanol	0.571
2-butanol	0.447
2-propanol	0.799
Dichloromethane	0.042
Formic acid	0.722
Acetic acid	0.174
Ethylene glycol	1.350
Acetone	0.054
Toluene	0.040

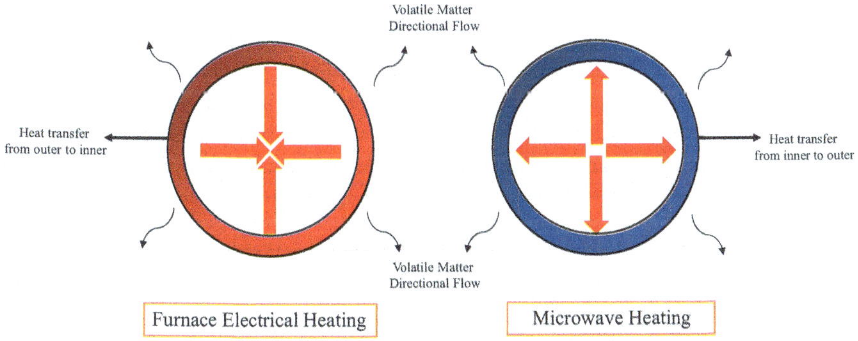

Fig. 3 Heat flow of furnace electrical heating and microwave heating [105, 108]

materials as the heating process is slower to penetrate the inner core region. These issues can be overcome by microwave heating.

3.3 Fast Microwave-Assisted Pyrolysis

Fast microwave-assisted pyrolysis reactor setup can be complex compared to other conventional pyrolysis system. Based on several research work [78, 106], the current fast microwave-assisted pyrolysis can be mainly in two forms, pre-mixed system (batch system) or post-mixed system (feeder system) (Fig. 3). For batch system, the main difference is that the reactor is placed in the microwave without a feeder inlet. Usually, for microwave reactor, a thermocouple is needed to monitor the temperature ramp caused by heating thereby allowing a control on the power to increase or reduce its intensity. To ensure the effective penetration of microwave irradiation, quartz materials which can withstand high temperature are often selected because of their high microwave penetration depth (d_P). However, despite so, the high cost associated with quartz reactor is also undesirable. Thus, a more common material such as borosilicate glass is used as an alternative.

For post-mix system, the catalyst (if any) will be mixed with the adsorbent in the microwave bed before introducing the waste materials through the feeder. This is so ensure that when the wastes are pyrolysed by microwave, the effective heating is imposed on the waste materials leading to shorter pyrolysis time and lower temperature gradient. Unlike conventional pyrolysis control, fast microwave-assisted pyrolysis can either control the temperature and power of the pyrolysis conditions, instead of heating rates. This is because the physicochemical and biochemical of the wastes change their tan δ values, which thereby affecting the heating rates during the pyrolysis process. If more water molecules are presented in the wastes, this might accelerate the heat conversion. However, too much water molecules in the waste can also lead to undesirable generation of low-quality bio-oil. Therefore, this should be balanced to optimise the heating without jeopardising the product's quality. Once the pyrolysis process is initiated, the evolution of volatiles will be channelled to a series of condensers to quickly quench the bio-oil into the downstream of the collectors.

The combination of pyrolysis and microwave system can lead to many parametric control factors, as per the following summary [13, 103, 106, 110]:

- Pyrolysis setting: Heating rate; heating temperature; residence time;
- Feedstock: Biochemical contents; physicocwever, from this particular hemical properties; particle size;
- Microwave setting: Output power; Type of mode (multimode or single-mode), receptor conditions (absorbability, concentration), pre-mixed or non-pre-mixed of absorbent/catalyst
- Addition setting: Use of catalyst/additives.

As shown in Fig. 4 as general setting, fast microwave-assisted pyrolysis, requires short pre-op time for start-up and shut down [107]. This is an attractive technology

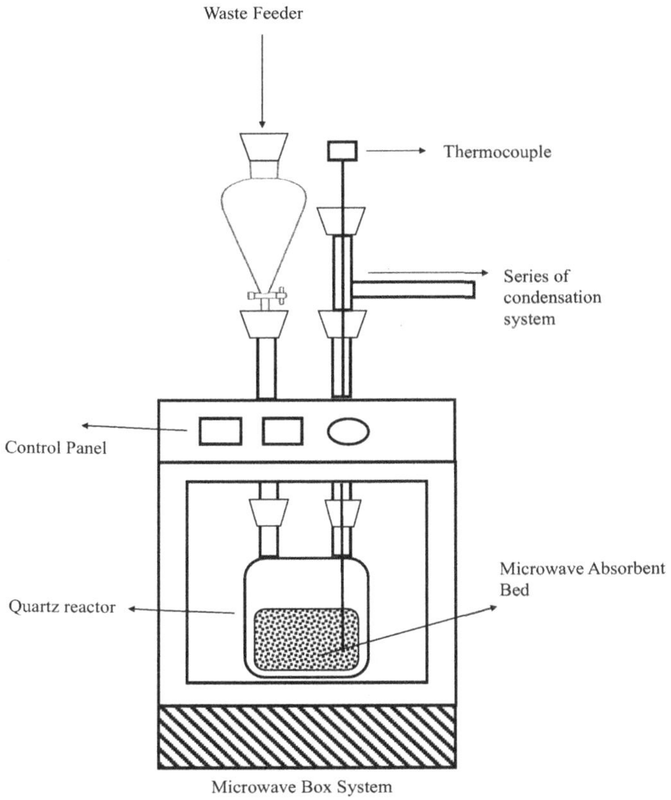

Fig. 4 General schematic for fast microwave-assisted pyrolysis

which can reduce the downtime for maintenance and repair. Furthermore, it is said that this technology produces bio-oil with lower ash content, purer syngas and can be scaled up more easily.

4 Waste Materials as Feedstock for Fast Microwave-Assisted Pyrolysis

As discussed in Sect. 2, biofuels are standardised based on their feedstock characteristics. However, the categorisation waste materials remain to be challenging due to their different chemical and biochemical contents. To date, fast microwave-assisted pyrolysis has been gaining more focuses on different waste materials, such as agricultural residues, wood residues, sludge wastes, waste tyres, plastic wastes as well as other organic wastes [111]. In this section, the wastes will be categorised into lignocellulosic-based materials (mainly consist of holocellulose, extractives, lignin and ash) and non-lignocellulosic-based materials (proteins, lipids, inorganics, minerals and others) [3].

4.1 *Lignocellulosic Wastes*

4.1.1 Agricultural Wastes

Agricultural wastes or residues are by far one of the most generated biomass in the world [112]. Depending on the agricultural activities on the respective countries, the generated agricultural wastes can be very different. Nonetheless, some of the more common biomass wastes produced from agricultural sectors, include corn stover [113], barley straw [113], oil palm wastes [6, 114], rice husk [115, 116] as well as rice straw [117], and many more. These materials are also rich with holocellulose and lignin which both have high microwave absorbability (0.035–0.062) as shown in Table 4.

Omar et al. [114] investigated the fast microwave-assisted pyrolysis on empty fruit bunch (EFB). The study investigated in-depth on the dielectric constants of the EFB at 2.45 GHz, and found to be proportional to its moisture content. However, the research work did not investigate parametric control on the EFB pyrolysis. In another study, *Vitis vinifera* was pyrolysed under the presence of microwave absorber (Fe), resulting in low viscosity (1.88–2.33 cP) and high yield (34.9%) of bio-oils [118]. From the same study, the use of a fractionation column in the outlet system generated higher yield of gas portion (45.7 wt%) using SiC as the absorber. Bouxin et al. [113] combined the fast microwave-assisted pyrolysis system with steam distillation, and applied this new technology on barley wastes. It was found that the levoglucosan and levoglucosene were the dominant products for barley feedstock treated with less than 0.8 wt% of acid. Thus, it can be observed that the experimental setup also greatly influenced the selectivity of the compounds in the biofuels aside from their yields.

In China and other Asian countries, rice is the main food products, which results in a significant amount of residues such as rice straw and husks on a yearly basis burdening the landfill system [116, 117]. Instead of disposing these wastes, they are sources of untapped and potential bioenergy readily available. However, the rice wastes as agricultural wastes have not been thoroughly studied as feedstock for microwave-assisted pyrolysis until in recent years. For instance, a new continuous fast microwave-pyrolysis system was studied on rice straw and *Camellia oleifera*, respectively at different feed rates, pyrolysis temperature on the optimisation of bio-oil production [117]. From the study, at increasing temperature, the bio-oil yield produced from rice husk increased accordingly; but for *Camellia oleifera*, the yield decreased with increasing temperature. This can be explained that both materials have very different lignocellulosic contents, which directly affect their thermal degradation profiles and trends. To enhance the pyrolysis process and to improve the selectivity, catalysts are often in this case. The following two studies applied alkali-treated

HZSM-5 zeolite [116] and hierarchical HZSM-5/MCM-41 catalyst [115], respectively on rice husk. These catalysts not only promote the reactivity of the pyrolysis process, and also improve the yield and selectivity of the targeted compounds.

4.1.2 Wood and Plant Wastes

For many years, wood and plant wastes are generated constantly from logging activities and deforestation. Similar to agricultural wastes, wood and plant wastes are generated throughout the year, and reforestation can ensure its sustainable supplies as feedstock for fast microwave-assisted pyrolysis. Some of the more common wood wastes are wood sawdust, wood chips, pine sawdust, palm tree trunk, rubber trees and others [119]. To date, there are many different soft wood and hard wood species available in the nature, whereby lignin is usually their main constituent in the structure which gives them the physical strength [120]. For pyrolysis activation, Ding et al. [120] found that in their studies, the selected softwood (*Cunninghamia lanceolata*) showed high activation energies compared to hardwood (*Fagus sylvatica*), suggesting that hardwood in this case was more prone to thermal degradation. Nonetheless, this also depends on their physicochemical and biochemical contents.

Borges et al. [107] investigated the use of wood sawdust as feedstock for fast microwave-assisted pyrolysis process. From result analysis, the bio-oil from wood sawdust yielded at 65 wt% with the optimum temperature at 480 °C with 2 g min^{-1} of feeding rate. Plant waste such as soapstock was also used as feedstock for fast microwave-assisted pyrolysis [121]. The yield of the hydrocarbon fuel from soapstock reached up to 64.74 wt%, with high amounts of alkanes and alkanes and higher calorific values compared to biodiesel.

As lignocellulosic wastes, wood sawdust is not rich in biochemical such as proteins and lipids. Therefore in some pyrolysis strategies, fast microwave-assisted co-pyrolysis can be applied. By mixing with *Chlorella vulagris* with wood sawdust, this could introduce high protein, carbohydrate and lipid contents into the pyrolysis process [122]. However, from this particular study of mixing algae into the wood sawdust, the calorific value of the bio-oil decreased after mixing both lignocellulosic and non-lignocellulosic wastes for pyrolysis. Another study also attempted the fast microwave-assisted co-pyrolysis on four Mediterranean available forest wastes such as *Pinus pinaster*, *Quercus ilex*, pinewoods and *Cistus ladanifer* [123]. These wastes were considered to be fully lignocellulosic. By varying the microwave power from 300 to 400 W, the bio-oil yield increased about 13–58%. Therefore, observing from the current literature results, co-pyrolysis can be quite complex whereby the compatibility between the biochemical contents and lignocellulosic fractions of the waste materials might be difficult to predict. Thus, such co-pyrolysis strategy requires more experimental development to form the fundamentals of compatibility system for lignocellulosic and non-lignocellulosic materials.

5 Non-lignocellulosic Wastes

5.1 Sludge Waste

Sludge waste are produced from both municipal wastewater and industrial wastewater treatments, which eventually will end up in landfill. Generally, sludge composition is formed from the degradation of different wastes rich with proteins, carbohydrates, lipids, and usually high ash contents (>10%) [108, 124]. Sludge wastes are mainly made up of 60% non-toxic organic materials, some minerals, and water but can be tainted with toxic metals, organic pollutants and pathogens which cannot be disposed directly [108]. Some of the more concerning sludge wastes are originated from textile factories, oil and gas industries and electroplating factories. Therefore, fast microwave-assisted pyrolysis can impartially treat sludge waste as feasible feedstock for biofuels production.

Commonly, sewage sludge is generated abundantly in the city area from municipal sources, causing certain problem in waste management. To resolve such an issue, Xie et al. [78] collected sewage sludge and subjected this waste with fast microwave-assisted pyrolysis with the aid of HZSM-5 catalyst. The bio-oil yield was negatively affected by the presence of the catalyst; however, its quality was improved with lower oxygen- and nitrogen-compounds presence in the liquid portion. In another study, pulp mill sludge was optimised of its solid portion by using fast microwave-assisted pyrolysis [125]. About 5% KOH was added to the pyrolysis reaction as microwave absorber, which in return, generated high-quality activated carbon in mere three minutes reaction.

Textile sludge can be quite difficult to handle, due to its hazardous and complex nature [126]. Furthermore, the composition of the sludge wastes vary from batch to batch, which further complicates and burdens the treatment process. Thus, an auger pyrolyser equipped with microwave heating system was used to convert the textile sludge into pyrolytic products [126]. In the bio-oil portion, the pyrolysis temperature played a role in controlling the aromatic compounds. The presence of additives (CaO, Fe, Ca-bentonite) has multiple roles such as microwave absorbers, catalysts, sorbents as well as reactants in this case.

Oil-based sludge wastes have been a longstanding issues to the environment, as they are very recalcitrant to natural degradation due to their compositions including oil (up to 80%), water (up to 50%), solid particle or soil (up to 20%), along with other chemical compounds such as stabiliser, surfactant and toxic metals [127]. For the following study [127], a magnetron of 1.5 kW and 2.45 GHz was used in the microwave-pyrolysis reactor as the heating element for the fast pyrolysis of oil sludge originated for crude oil extraction platform in China. The use of catalyst (β-zeolite) in the process improved the bio-oil quality by lowering oxygenates in its composition. At 500 °C, the highest oil yield could be obtained was 85.93 wt% with only 15 min of pyrolysis time. From another research work, the fast microwave-assisted pyrolysis on two different oil sludge, namely sludge 1 (volatile matter and hydrocarbon dominant) and sludge 2 (ash, nitrogen, sulphur and iron dominant). Both showed excellent bio-oil yields up to 87.3 wt% and 41.2 wt%, respectively. These oil sludge-based bio-oils were further upgraded through ex situ catalytic system, resulting in significant reduction of deoxygenates (<5%) with great calorific values (44–45 MJ kg^{-1}). From the current studies, oil sludge showed high yields in liquid portion and showed great potential to be upgraded as biofuels directly. Based on the current research development, much of the work need to be done on establishing fast microwave-assisted pyrolysis on sludge wastes. Due to the compositional varieties of these sludge materials, their pyrolysis mechanisms might differ from the general pyrolysis zones. This would post great challenges in setting up a commercial pyrolyser without these critical information.

5.2 *Polymer Waste*

Polymer wastes including plastics and natural rubbers are highly carbon rich materials which contain organic compounds. So far, different plastic wastes such as polypropylene (PP), high-density polyethylene (HDPE), low-density polyethylene (LDPE), and polystyrene (PS), as well as waste tyres have been investigated as feedstock for fast microwave-assisted pyrolysis [105].

As observed from Table 4, plastics possess very low tan δ values (0.0001–0.0056) which mean that they might not be effectively subjected to microwave heating. However, plastics as fossil-based materials, contain many carbon chains available for conversion into fuel products. Since plastics are not thermally activated for thermal degradation state, the pyrolysis heating would not be feasible to form biofuels. With such shortcoming with plastic waste feedstock, fast microwave-assisted pyrolysis can be one of the solutions to be applied in the form of co-pyrolysis. In the following study, LDPE was co-pyrolysed with lignin under the catalytic reaction with HZSM-5 and MgO [128]. The synergistic interaction between LDPE and lignin was also evaluated based on the theoretical and experimental bio-oil yields. From the results, LDPE was identified to be a good hydrogen donor thereby improving the bio-oil quality. Catalytic effects of MZSM-5 improved the aromatics generation while MgO enhances the phenols' alkylation. Similar approach was also used on the co-pyrolysis of distillers dried grains with soluble and plastic mulching films [129]. In this study, the ratio of two feedstock was investigated and found to have great impacts on the bio-oil yield and the inhibition on the coke formation. Subsequently for the following fast microwave-assisted co-pyrolysis, the combination of PP and bamboo (bio-oil yield: 61.62 wt%) [130], rice straw and PP (bio-oil yield: 31.5 wt%), rice straw and PS (bio-oil yield: 43.0 wt%), sugarcane bagasse and PP (bio-oil yield: 39.3 wt%) sugarcane bagasse and PS (bio-oil yield: 56.5 wt%) [131] all showed high yield of bio-oil fractions.

As for rubber wastes, their tan δ values range in between 0.002 and 0.005 which are slightly better microwave absorber compared to plastic materials. Usually, rubber wastes are originated mainly from automobile industry in the form of waste tyres

[132, 133]. With the global development in automobile industry, it is anticipated more and more tyres will be disposed over the course of time. As the tyres are embedded with steel wires, compression on them are hardly effective and disposal to landfill can easily wear out the capacity. Thus, fast microwave-assisted pyrolysis not only can resolve the waste tyres in a rapid manner, this technology can also help to recover valuable commodities such as steel, carbon and biofuels. To pyrolise waste tyres, the feedstock must be fragmented to suitable sizing for the feeding to reactor. One of the studies investigated the fast microwave-assisted pyrolysis on two commercially available tyres, with different compositions [132]. From the results, it was found that tyres with higher styrene contents were converted more rapidly to pyrolytic phases compared to natural rubber bounded wastes. Nonetheless, both tyre wastes generated high bio-oil yield (>40 wt%), with low densities (0.874–0.921 g cm^{-3}) and low viscosities (1.73–3.92 cP), suitable for biofuel upgrade. Co-pyrolysis approach was also applied on several interesting combinations in this case to enhance the biofuels production. Waste tyre was co-pyrolysed with soapstock [134], bamboo sawdust [135] and chlorella [136]. To optimise the bio-oil fraction, the ratio of both feedstock requires precise adjustment in order to achieve the set target. So far, the bio-oil yields from these studies ranged in between 30.23 and 52.67 wt%, which considered to be excellent. Unlike sludge wastes, the application of fast microwave-assisted pyrolysis needs to be focusing on the materials' microwave absorbability, which greatly impact the pyrolysis process. From the studies, fast microwave-assisted pyrolysis can be integrated with co-pyrolysis to overcome the current shortcomings.

6 Challenges and Conclusion

6.1 Waste Feedstock

Waste materials despite can be found almost in any countries depending on their origins, still face the issue of feedstock consistency. That being said, for bioenergy to be developed globally, this will remain as a major challenge, as not all countries produce their own wastes sufficiently to be used for energy generation. However, in some countries, the wastes are exceeding their landfill and disposal systems. Therefore, WtE or EfW needs to be a collaborative effort between countries. In fact, waste materials can be a form of commodity in the global trade in the future for energy generation feedstock. However, as these are waste materials, their generation qualities and quantities are usually not guaranteed, which would pose processing issue in the downstream of EfW facilities. From the current studies, the information on fast microwave-assisted pyrolysis of mixing wastes are growing, yet still lacking in the industrial scale at the moment. This needs to be addressed for future development of fast microwave-assisted pyrolysis specifically for biofuels, to establish a good ratio for co-pyrolysis as well as to ensure process stability given the feedstock's characteristics.

6.1.1 Technology Limitation

One of the greatest challenges associated with microwave technology is its limitation on frequency selection, to ensure no interference on other microwave system, fixing at 24.5 GHz. This undeniably has limited further development in perfecting the suitable system for pyrolysis process. However, for higher power application (>60 kW), the current microwave is sufficiently energy intensive. For more energy efficient and deeper wave penetration for waste materials, 915 MHz microwave system can be of great choice in large scale application [137]. However, at the moment, this form of microwave system is yet to be common nor well-researched specifically for fast microwave-assisted pyrolysis of waste materials.

Despite current 24.5 GHz microwave system is popular in electronics and food industry, the scaling up for pyrolysis system has not been further established due to the complex setup and safety concern during operation. Exposure to microwave irradiation can be harmful to human bodies, causing high degree of burn injuries and sometimes leading to cataracts. Therefore, to ensure the operation of a massive microwave system is safe, meticulous engineering designs and details are required during the design phase, commission stage and full operation.

7 Conclusions

The unhealthy energy mix across the world is a worrying trend, as fossil fuels are not only unsustainable, they also produce undesirable pollutants to the atmosphere. Thus, the share of the renewable energies should be increased from time to time, in order to achieve a balance energy mix in the near future. Bioenergy has huge potential to replace fossil fuels in a long run, especially with biofuels generated from wastes providing lower carbon footprint and comparable energy contents. The exploration on producing biofuels using fast microwave-assisted pyrolysis has been emerging and used on different lignocellulosic wastes (agricultural wastes, wood and plant wastes) and non-lignocellulosic wastes (sludge waste and polymer waste). From the literatures, each waste group possesses different pyrolysis behaviours and generalising these behaviours would be useful for future commercialisation. At times, these two categories of wastes can be applied in fast microwave-assisted co-pyrolysis to generate synergistic benefits for maximise the yield and quality of biofuels. Nonetheless, based on the current literature and industries practice, the development of fast microwave-assisted pyrolysis is still considered to be in a steady progress. As long as the identified challenges associated with waste feedstock and technology limitations can be resolved, fast microwave-assisted pyrolysis has the potential to be one of the core thermochemical technologies in producing biofuels from wastes in the energy sector.

References

1. Curley R (2012) Fossil fuels energy: past, present and future. Britannica Educational Planning, New York
2. IEA (2021) Source: IEA World Energy Balances 2020. In: International Energy Agency
3. Lee XJ, Ong HC, Gan YY, Chen W-H, Mahlia TMIJEC (2020) Management, State of art review on conventional and advanced pyrolysis of macroalgae and microalgae for biochar, bio-oil and bio-syngas production. Energy Convers Manage 210:112707
4. Crutzen PJ, Mosier AR, Smith KA, Winiwarter W (2016) N_2O release from agro-biofuel production negates global warming reduction by replacing fossil fuels. In: Crutzen PJ (ed) A pioneer on atmospheric chemistry and climate change in the anthropocene, Springer, pp 227–238
5. Peter SC (2018) Reduction of CO_2 to chemicals and fuels: a solution to global warming and energy crisis. ACS Energy Lett 3:1557–1561
6. Lee XJ, Lee LY, Gan S, Thangalazhy-Gopakumar S, Ng HK (2017) Biochar potential evaluation of palm oil wastes through slow pyrolysis: Thermochemical characterization and pyrolytic kinetic studies. Biores Technol 236:155–163
7. Ahmed A, Abu Bakar MS, Azad AK, Sukri RS, Mahlia TMI (2018) Potential thermochemical conversion of bioenergy from *Acacia* species in Brunei Darussalam: a review. Renew & Sustain Energy Rev 82:3060–3076
8. Lin J, Ma R, Luo J, Sun S, Cui C, Fang L, Huang H (2020) Microwave pyrolysis of food waste for high-quality syngas production: positive effects of a CO_2 reaction atmosphere and insights into the intrinsic reaction mechanisms. Energy Convers Manage 206:112490
9. Chanaka Udayanga WD, Veksha A, Giannis A, Lim TT (2019) Pyrolysis derived char from municipal and industrial sludge: impact of organic decomposition and inorganic accumulation on the fuel characteristics of char. Waste Manag 83:131–141
10. Gupta GK, Mondal MK (2022) Pyrolysis: an alternative approach for utilization of biomass into bioenergy generation. In: Biofuels and bioenergy. Elsevier, pp 279–300
11. Tripathi M, Sahu JN, Ganesan P (2016) Effect of process parameters on production of biochar from biomass waste through pyrolysis: A review. Renew Sustain Energy Rev 55:467–481
12. Li DC, Jiang H (2017) The thermochemical conversion of non-lignocellulosic biomass to form biochar: A review on characterizations and mechanism elucidation. Bioresour Technol 246:57–68
13. Lee XJ (2018) Evaluation of cost effective adsorbent and biochar from Malaysia oil palm wastes: synthesis, characterisation and optimisation studies. In Chemical and environmental engineering. University of Nottingham, United Kingdom
14. IEA (2021) Source: IEA Data Services. In: International Energy Agency
15. Shafiee S, Topal E (2009) When will fossil fuel reserves be diminished? Energy Policy 37:181–189
16. IEA (2021) Source: IEA Electricity Information 2020. In: International Energy Agency
17. Daly H (1994) Fossil fuels. Appl Energy 47:101–121
18. Casper JK (2010) Fossil fuels and pollution: the future of air quality. Facts On File, Inc., United States of America
19. Pate-Cornell ME (1993) Learning from the piper alpha accident: a postmortem analysis of technical and organizational factors. Risk Anal 13:215–215
20. Peterson CH (2001) The "Exxon Valdez" oil spill in Alaska: Acute, indirect and chronic effects on the ecosystem. In: Advances in marine biology, Academic Press, pp 1–103
21. Lilley J, Firestone J (2013) The effect of the 2010 Gulf oil spill on public attitudes toward offshore oil drilling and wind development. Energy Policy 62:90–98
22. Fantazzini D (2016) The oil price crash in 2014/15: was there a (negative) financial bubble? Energy Policy 96:383–396
23. White JE, Catallo WJ, Legendre BL (2011) Biomass pyrolysis kinetics: a comparative critical review with relevant agricultural residue case studies. J Anal Appl Pyrol 91:1–33

24. Zinkle SJ, Was GS (2013) Materials challenges in nuclear energy. Acta Mater 61:735–758
25. Li X, Wu J, Liao J, Zhang D, Yang J, Feng Y, Zeng J, Wen W, Yang Y, Tang J, Liu N (2013) Adsorption and desorption of uranium (VI) in aerated zone soil. J Environ Radioact 115:143–150
26. Soltanieh AA, Rezaeian M (2017) Nuclear and renewable energy in Iran: nuclear challenges and opportunities. Int J Hydrog Energy
27. Sakuma K, Kitamura A, Malins A, Kurikami H, Machida M, Mori K, Tada K, Kobayashi T, Tawara Y, Tosaka H (2017) Characteristics of radio-cesium transport and discharge between different basins near to the Fukushima Dai-ichi Nuclear Power Plant after heavy rainfall events. J Environ Radioact 169–170:137–150
28. Sakamoto S, Kawase Y (2016) Adsorption capacities of poly-gamma-glutamic acid and its sodium salt for cesium removal from radioactive wastewaters. J Environ Radioact 165:151–158
29. IAEA (2021) Nuclear Share of Electricity Generation in 2020. In: International Atomic Energy Agency
30. Paish O (2002) Small hydro power: technology and current status. Renew Sustain Energy Rev 6:537–556
31. Mishra S, Singal SK, Khatod DK (2011) Optimal installation of small hydropower plant—a review. Renew Sustain Energy Rev 15:3862–3869
32. Kishor N, Saini RP, Singh SP (2007) A review on hydropower plant models and control. Renew Sustain Energy Rev 11:776–796
33. Güney MS, Kaygusuz K (2010) Hydrokinetic energy conversion systems: a technology status review. Renew Sustain Energy Rev 14:2996–3004
34. Evans A, Strezov V, Evans TJ (2009) Assessment of sustainability indicators for renewable energy technologies. Renew Sustain Energy Rev 13:1082–1088
35. Rahman Mohamed A, Lee KT (2006) Energy for sustainable development in Malaysia: energy policy and alternative energy. Energy Policy 34:2388–2397
36. Wu J, Huang J, Han X, Gao X, He F, Jiang M, Jiang Z, Primack RB, Shen Z (2004) The Three Gorges Dam: an ecological perspective. Front Ecol Environ 2:241–248
37. Yang Z, Wang H, Saito Y, Milliman JD, Xu K, Qiao S, Shi G (2006) Dam impacts on the Changjiang (Yangtze) River sediment discharge to the sea: the past 55 years and after the Three Gorges Dam. Water Resour Res 42:n/a–n/a
38. Keong CY (2005) Energy demand, economic growth, and energy efficiency—the Bakun dam-induced sustainable energy policy revisited. Energy Policy 33:679–689
39. Okot DK (2013) Review of small hydropower technology. Renew Sustain Energy Rev 26:515–520
40. Ong HC, Mahlia TMI, Masjuki HH (2011) A review on energy scenario and sustainable energy in Malaysia. Renew Sustain Energy Rev 15:639–647
41. Chua SC, Oh TH, Goh WW (2011) Feed-in tariff outlook in Malaysia. Renew Sustain Energy Rev 15:705–712
42. Ackermann T, Söder L (2000) Wind energy technology and current status: a review. Renew Sustain Energy Rev 4:315–374
43. Price TJ (2005) James Blyth—Britain's first modern wind power pioneer. J Wind engineering 29:191–200
44. Şahin AD (2004) Progress and recent trends in wind energy. Prog Energy Combust Sci 30:501–543
45. Baños R, Manzano-Agugliaro F, Montoya FG, Gil C, Alcayde A, Gómez J (2011) Optimization methods applied to renewable and sustainable energy: A review. Renew Sustain Energy Rev 15:1753–1766
46. Joselin Herbert GM, Iniyan S, Sreevalsan E, Rajapandian S (2007) A review of wind energy technologies. Renew Sustain Energy Rev 11:1117–1145
47. Ackermann T, Söder L (2002) An overview of wind energy-status 2002. Renew Sustain Energy Rev 6:67–127

48. Ahmad S, Kadir MZAA, Shafie S (2011) Current perspective of the renewable energy development in Malaysia. Renew Sustain Energy Rev 15:897–904
49. C.S. Solanki, Solar photovoltaics: fundamentals, technologies and applications, Phi learning pvt. Ltd., 2015.
50. Guerrero-Lemus R, Vega R, Kim T, Kimm A, Shephard LE (2016) Bifacial solar photovoltaics—a technology review. Renew Sustain Energy Rev 60:1533–1549
51. Ahmad AA, Migdadi AB, Alsaad AM, Al-Bataineh QM, Telfah A (2021) Optical, structural, and morphological characterizations of synthesized (Cd–Ni) co-doped ZnO thin films. Appl Phys A 127:922
52. Wang S, Huang L, Zhong L, Zhi Z, Li J, Xiao X (2021) Rear interface modification by the ZnTe layer enables high-efficient $Cu_2(Zn,Cd)SnS_4$ thin-film solar cells. ACS Appl Energy Mater
53. Cavalli A, Alkurd N, Johnston S, Diercks DR, Roberts DM, Ley BE, Simon J, Young DL, Packard CE, Ptak AJJIJOP (2021) Performance of III–V solar cells grown on reformed mesoporous Ge templates
54. Caño P, Hinojosa M, García I, Beanland R, Marrón DF, Ruiz CM, Johnson A, Rey-Stolle IJSE (2021) GaAsP/SiGe tandem solar cells on porous Si substrates. Solar Energy 230:925–934
55. Wang K, Zheng L, Zhu T, Liu L, Becker ML, Gong XJNE (2020) High performance perovskites solar cells by hybrid perovskites co-crystallized with poly (ethylene oxide). Nano Energy 67:104229
56. Bhatia SC (2014) 4—solar thermal energy. In: Bhatia SC (ed) Advanced renewable energy systems. Woodhead Publishing India, pp 94–143
57. Schwede JW, Bargatin I, Riley DC, Hardin BE, Rosenthal SJ, Sun Y, Schmitt F, Pianetta P, Howe RT, Shen Z-X, Melosh NA (2010) Photon-enhanced thermionic emission for solar concentrator systems. Nat Mater 9:762–767
58. Bahaj AS, Blunden LS (2007) Tidal energy resource assessment for tidal stream generators. Proc Inst Mech Eng Part A: J Power Energy 221:137–146
59. Rourke FO, Boyle F, Reynolds A (2010) Tidal energy update 2009. Appl Energy 87:398–409
60. Laughton MA (2003) Renewable energy sources watt committee report number 22. CRC Press Taylor & Francis Group, England
61. Denny E (2009) The economics of tidal energy. Energy Policy 37:1914–1924
62. Westwood A (2004) Ocean power. Refocus 5:50–55
63. Barbier E (2002) Geothermal energy technology and current status: an overview. Renew Sustain Energy Rev 6:3–65
64. Barbier E (1997) Nature and technology of geothermal energy: A review. Renew Sustain Energy Rev 1:1–69
65. Fridleifsson IB (2001) Geothermal energy for the benefit of the people. Renew Sustain Energy Rev 5:299–312
66. Lund JW, Freeston DH, Boyd TL (2011) Direct utilization of geothermal energy 2010 worldwide review. Geothermics 40:159–180
67. Lund JW, Freeston DH, Boyd TL (2005) Direct application of geothermal energy: 2005 Worldwide review. Geothermics 34:691–727
68. Rybach L (2003) Geothermal energy: sustainability and the environment. Geothermics 32:463–470
69. Johari A, Ahmed SI, Hashim H, Alkali H, Ramli M (2012) Economic and environmental benefits of landfill gas from municipal solid waste in Malaysia. Renew Sustain Energy Rev 16:2907–2912
70. Rodionova MV, Poudyal RS, Tiwari I, Voloshin RA, Zharmukhamedov SK, Nam HG, Zayadan BK, Bruce BD, Hou HJ, Allakhverdiev SIJIJOHE (2017) Biofuel production: challenges and opportunities. Int J Hydrog Energy 42:8450–8461
71. Srivastava RK, Shetti NP, Reddy KR, Aminabhavi TMJECL (2020) Biofuels, biodiesel and biohydrogen production using bioprocesses. A review. Environ Chem Lett 18:1049–1072
72. Sharma YC, Singh B (2009) Development of biodiesel: Current scenario. Renew Sustain Energy Rev 13:1646–1651

73. Puricelli S, Cardellini G, Casadei S, Faedo D, Van den Oever A, Grosso MJR, Reviews SE (2021) A review on biofuels for light-duty vehicles in Europe. Renew Sustain Energy Rev 137:110398
74. Lee XJ, Lee LY, Hiew BYZ, Gan S, Thangalazhy-Gopakumar S, Ng HK (2020) Valorisation of oil palm wastes into high yield and energy content biochars via slow pyrolysis: multivariate process optimisation and combustion kinetic studies. Materials Science for Energy Technologies 3:601–610
75. Lee XJ, Ong HC, Gao W, Ok YS, Chen W-H, Goh BHH, Chong CT (2021) Solid biofuel production from spent coffee ground wastes: process optimisation, characterisation and kinetic studies. Fuel 292:120309
76. Dai L, Wang Y, Liu Y, Ruan R, Duan D, Zhao Y, Yu Z, Jiang L (2019) Catalytic fast pyrolysis of torrefied corn cob to aromatic hydrocarbons over Ni-modified hierarchical ZSM-5 catalyst. Bioresour Technol 272:407–414
77. Saxena RC, Adhikari DK, Goyal HB (2009) Biomass-based energy fuel through biochemical routes: a review. Renew Sustain Energy Rev 13:167–178
78. Xie Q, Addy M, Liu S, Zhang B, Cheng Y, Wan Y, Li Y, Liu Y, Lin X, Chen P, Ruan R (2015) Fast microwave-assisted catalytic co-pyrolysis of microalgae and scum for bio-oil production. Fuel 160:577–582
79. Mirza UK, Ahmad N, Majeed T (2008) An overview of biomass energy utilization in Pakistan. Renew Sustain Energy Rev 12:1988–1996
80. Kan T, Strezov V, Evans TJ (2016) Lignocellulosic biomass pyrolysis: A review of product properties and effects of pyrolysis parameters. Renew Sustain Energy Rev 57:1126–1140
81. Bruckner T, Chum H, Jäger-Waldau A, Killingtveit Å, Gutiérrez-Negrín L, Nyboer J, Musial W, Verbruggen A, Wiser R (2011) Annex III: recent renewable energy cost and peformance parameters. In: Edenhofer RP-MO, Sokona Y, Seyboth K, Matschoss P, Kadner S, Zwickel T, Eickemeier P, Hansen SSG, von Stechow C (eds) IPCC special report on renewable energy sources and climate change mitigation. Cambridge University Press, Cambridge, United Kingdom and New York, NY, USA
82. IEA (2020) Projected costs of generating electricity 2020. In International Energy Agency.
83. Hienuki S, Mitoma H, Ogata M, Uchida I, Kagawa SJIJOHE (2021) Environmental and energy life cycle analyses of passenger vehicle systems using fossil fuel-derived hydrogen. Int J Hydrog Energy 46:36569–36580
84. Krebs-Moberg M, Pitz M, Dorsette TL, Gheewala SHJRE (2021) Third generation of photovoltaic panels: a life cycle assessment. Renew Energy 164:556–565
85. Teffera B, Assefa B, Björklund A, Assefa GJTIJOLCA (2021) Life cycle assessment of wind farms in Ethiopia. Int J Life Cycle Assess 26:76–96
86. Sigurjónsson HÆ, Cook D, Davíðsdóttir B, Bogason SGJRE (2021) A life-cycle analysis of deep enhanced geothermal systems–the case studies of Reykjanes, Iceland and Vendenheim, France. Renew Energy 177:1076–1086
87. Elsamadony M, Elreedy A, Mostafa A, Fujii M, Gescher J, Shakeri Yekta S, Schnurer A, Gaillard J-F, Pant DJASC (2021) Engineering, perspectives on potential applications of nanometal derivatives in gaseous bioenergy pathways: mechanisms, life cycle, and toxicity. ACS Sustain Chem Eng 9:9563–9589
88. Schlömer S, Bruckner T, Fulton L, Hertwich E, McKinnon A, Perczyk D, Roy J, Schaeffer R, Sims R, Smith P, Wiser R (2014) Annex III: technology-specific cost and performance parameters. In Edenhofer O, Pichs-Madruga R, Sokona Y, Farahani E, Kadner S, Seyboth K, Adler A, Baum I, Brunner SP, Eickemeier BK, Savolainen J, Schlömer S, von Stechow C, Zwickel T, Minx JC (eds) Climate change 2014: mitigation of climate change. Contribution of working group III to the fifth assessment report of the intergovernmental panel on climate change. Cambridge University Press, Cambridge, United Kingdom and New York, NY, USA
89. Yu J, Maliutina K, Tahmasebi A (2018) A review on the production of nitrogen-containing compounds from microalgal biomass via pyrolysis. Bioresour Technol 270:689–701
90. Lee XJ, Ong HC, Ooi J, Yu KL, Tham TC, Chen W-H, Ok YS (2022) Engineered macroalgal and microalgal adsorbents: Synthesis routes and adsorptive performance on hazardous water contaminants. J Hazard Mater 423:126921

91. Demirbas A (2009) Pyrolysis mechanisms of biomass materials. Energy Sources, Part A: Recover Utilization Environ Effects 31:1186-1193.
92. Demirbaş A (2000) Mechanisms of liquefaction and pyrolysis reactions of biomass. Energy Convers Manage 41:633–646
93. Van de Velden M, Baeyens J, Brems A, Janssens B, Dewil R (2010) Fundamentals, kinetics and endothermicity of the biomass pyrolysis reaction. Renew Energy 35:232–242
94. Shafizadeh F (1982) Introduction to pyrolysis of biomass. J Anal Appl Pyrol 3:283–305
95. Boukis I (1997) Fast pyrolysis of biomass in a circulating fluidized bed reactor. Aston University, United Kingdom
96. Raveendran K, Ganesh A, Khilar KC (1996) Pyrolysis characteristics of biomass and biomass components. Fuel 75:987–998
97. Bach QV, Chen WH (2017) Pyrolysis characteristics and kinetics of microalgae via thermogravimetric analysis (TGA): a state-of-the-art review. Bioresour Technol 246:88–100
98. Remya N, Lin J-G (2011) Current status of microwave application in wastewater treatment—a review. Chem Eng J 166:797–813
99. Li Z, Wang R, Kudra T (2011) Uniformity issue in microwave drying. Drying Technol 29:652–660
100. Cameron RJ, Kudsia CM, Mansour RR (2018), Microwave filters for communication systems: fundamentals, design, and applications. Wiley.
101. Yin C (2012) Microwave-assisted pyrolysis of biomass for liquid biofuels production. Bioresour Technol 120:273–284
102. Du Z, Li Y, Wang X, Wan Y, Chen Q, Wang C, Lin X, Liu Y, Chen P, Ruan R (2011) Microwave-assisted pyrolysis of microalgae for biofuel production. Bioresour Technol 102:4890–4896
103. Motasemi F, Afzal MT (2013) A review on the microwave-assisted pyrolysis technique. Renew Sustain Energy Rev 28:317–330
104. Zhang X, Rajagopalan K, Lei H, Ruan R, Sharma BK (2017) An overview of a novel concept in biomass pyrolysis: microwave irradiation. Sustainable Energy Fuels 1:1664–1699
105. Zhang Y, Chen P, Liu S, Peng P, Min M, Cheng Y, Anderson E, Zhou N, Fan L, Liu C, Chen G, Liu Y, Lei H, Li B, Ruan R (2017) Effects of feedstock characteristics on microwave-assisted pyrolysis—a review. Bioresour Technol 230:143–151
106. Zhang Y, Cui Y, Liu S, Fan L, Zhou N, Peng P, Wang Y, Guo F, Min M, Cheng Y, Liu Y, Lei H, Chen P, Li B, Ruan R (2020) Fast microwave-assisted pyrolysis of wastes for biofuels production—a review. Bioresour Technol 297:122480
107. Borges FC, Xie Q, Min M, Muniz LA, Farenzena M, Trierweiler JO, Chen P, Ruan R (2014) Fast microwave-assisted pyrolysis of microalgae using microwave absorbent and HZSM-5 catalyst. Bioresour Technol 166:518–526
108. Zaker A, Chen Z, Wang X, Zhang Q (2019) Microwave-assisted pyrolysis of sewage sludge: a review. Fuel Process Technol 187:84–104
109. Saber M, Nakhshiniev B, Yoshikawa K (2016) A review of production and upgrading of algal bio-oil. Renew Sustain Energy Rev 58:918–930
110. Fodah AEM, Ghosal MK, Behera D (2021) Microwave-assisted pyrolysis of agricultural residues: current scenario, challenges, and future direction. Int J Environ Sci Technol
111. Wang L, Lei H, Ruan R (2015) Techno-economic analysis of microwave-assisted pyrolysis for production of biofuels. In: Production of biofuels and chemicals with microwave, pp 251–263
112. Ravikumar C, Senthil Kumar P, Subhashni SK, Tejaswini PV, Varshini V, Microwave assisted fast pyrolysis of corn cob, corn stover, saw dust and rice straw: experimental investigation on bio-oil yield and high heating values. Sustain Mater Technol 11:19–27
113. Bouxin FP, Clark JH, Fan J, Budarin V (2019) Combining steam distillation with microwave-assisted pyrolysis to maximise direct production of levoglucosenone from agricultural wastes. Green Chem 21:1282–1291
114. Omar R, Idris A, Yunus R, Khalid K, Aida Isma MI (2011) Characterization of empty fruit bunch for microwave-assisted pyrolysis. Fuel 90: 1536–1544
115. Li Z, Zhong Z, Zhang B, Wang W, Zhao H, Seufitelli GVS, Resende FLP (2021) Microwave-assisted catalytic fast pyrolysis of rice husk over a hierarchical HZSM-5/MCM-41 catalyst prepared by organic base alkaline solutions. Sci Total Environ 750:141215

116. Li Z, Zhong Z, Zhang B, Wang W, Seufitelli GVS, Resende FLP (2020) Effect of alkali-treated HZSM-5 zeolite on the production of aromatic hydrocarbons from microwave assisted catalytic fast pyrolysis (MACFP) of rice husk. Sci Total Environ 703:134605
117. Wang Y, Zeng Z, Tian X, Dai L, Jiang L, Zhang S, Wu Q, Wen P, Fu G, Liu Y, Ruan R (2018) Production of bio-oil from agricultural waste by using a continuous fast microwave pyrolysis system. Bioresour Technol 269:162–168
118. Bartoli M, Rosi L, Giovannelli A, Frediani P, Passaponti M, Frediani M (2018) Microwave assisted pyrolysis of crop residues from *Vitis vinifera*. J Anal Appl Pyrol 130:305–313
119. Mushtaq F, Mat R, Ani FN (2014) A review on microwave assisted pyrolysis of coal and biomass for fuel production. Renew Sustain Energy Rev 39:555–574
120. Ding Y, Ezekoye OA, Lu S, Wang C, Zhou R (2017) Comparative pyrolysis behaviors and reaction mechanisms of hardwood and softwood. Energy Convers Manage 132:102–109
121. Wang Y, Dai L, Wang R, Fan L, Liu Y, Xie Q, Ruan R (2016) Hydrocarbon fuel production from soapstock through fast microwave-assisted pyrolysis using microwave absorbent. J Anal Appl Pyrol 119:251–258
122. Chen L, Yu Z, Xu H, Wan K, Liao Y, Ma X (2019) Microwave-assisted co-pyrolysis of Chlorella vulgaris and wood sawdust using different additives. Bioresour Technol 273:34–39
123. Martín MT, Sanz AB, Nozal L, Castro F, Alonso R, Aguirre JL, González SD, Matía MP, Novella JL, Peinado M, Vaquero JJ (2017) Microwave-assisted pyrolysis of Mediterranean forest biomass waste: bioproduct characterization. J Anal Appl Pyrol 127:278–285
124. Lee XJ, Lee LY, Hiew BYZ, Gan S, Thangalazhy-Gopakumar S, Kiat Ng H (2017) Multistage optimizations of slow pyrolysis synthesis of biochar from palm oil sludge for adsorption of lead. Bioresour Technol 245:944–953
125. Namazi AB, Grant Allen D, Jia CQ (2015) Microwave-assisted pyrolysis and activation of pulp mill sludge. Biomass Bioenergy 73:217–224
126. Zhang H, Gao Z, Ao W, Li J, Liu G, Fu J, Ran C, Mao X, Kang Q, Liu Y, Dai J (2017) Microwave-assisted pyrolysis of textile dyeing sludge using different additives. J Anal Appl Pyrol 127:140–149
127. Liu Y, Song Y, Zhang T, Jiang Z, Siyal AA, Dai J, Fu J, Zhou C, Wang L, Li X, Ao W, Jin X, Teng D, Fang J (2021) Microwave-assisted pyrolysis of oily sludge from offshore oilfield for recovery of high-quality products. J Hazard Mater 420:126578
128. Fan L, Chen P, Zhang Y, Liu S, Liu Y, Wang Y, Dai L, Ruan R (2017) Fast microwave-assisted catalytic co-pyrolysis of lignin and low-density polyethylene with HZSM-5 and MgO for improved bio-oil yield and quality. Bioresour Technol 225:199–205
129. Zhang B, Zhong Z, Li T, Xue Z, Wang X, Ruan R (2018) Biofuel production from distillers dried grains with solubles (DDGS) co-fed with waste agricultural plastic mulching films via microwave-assisted catalytic fast pyrolysis using microwave absorbent and hierarchical ZSM-5/MCM-41 catalyst. J Anal Appl Pyrol 130:1–7
130. Zhao Y, Wang Y, Duan D, Ruan R, Fan L, Zhou Y, Dai L, Lv J, Liu Y (2018) Fast microwave-assisted ex-catalytic co-pyrolysis of bamboo and polypropylene for bio-oil production. Bioresour Technol 249:69–75
131. Suriapparao DV, Vinu R, Shukla A, Haldar S (2020) Effective deoxygenation for the production of liquid biofuels via microwave assisted co-pyrolysis of agro residues and waste plastics combined with catalytic upgradation. Bioresour Technol 302:122775
132. Undri A, Rosi L, Frediani M, Frediani P (2014) Upgraded fuel from microwave assisted pyrolysis of waste tire. Fuel 115:600–608
133. Undri A, Sacchi B, Cantisani E, Toccafondi N, Rosi L, Frediani M, Frediani P (2013) Carbon from microwave assisted pyrolysis of waste tires. J Anal Appl Pyrol 104:396–404
134. Dai L, Fan L, Duan D, Ruan R, Wang Y, Liu Y, Zhou Y, Zhao Y, Yu Z (2017) Microwave-assisted catalytic fast co-pyrolysis of soapstock and waste tire for bio-oil production. J Anal Appl Pyrol 125:304–309
135. Wang Y, Dai L, Fan L, Duan D, Liu Y, Ruan R, Yu Z, Liu Y, Jiang L (2017) Microwave-assisted catalytic fast co-pyrolysis of bamboo sawdust and waste tire for bio-oil production. J Anal Appl Pyrol 123:224–228

136. Fang S, Gu W, Dai M, Xu J, Yu Z, Lin Y, Chen J, Ma X (2018) A study on microwave-assisted fast co-pyrolysis of chlorella and tire in the N_2 and CO_2 atmospheres. Bioresour Technol 250:821–827
137. Tyagi VK, Lo S-L (2013) Microwave irradiation: a sustainable way for sludge treatment and resource recovery. Renew Sustain Energy Rev 18:288–305

The manufacturer's authorised representative in the EU is Springer Nature Customer Service Centre GmbH, Europaplatz 3, 69115 Heidelberg, Germany. If you have any concerns regarding our products, please contact ProductSafety@springernature.com

Printed and bound by CPI Group (UK) Ltd, Croydon, CR0 4YY
01/10/2026
02225875-0005